T0128805

The Yucatan Hall of Records

The Yucatan Hall of Records:

Chichen Itza's Time Traveler Cult

Vernon Q. Murray, Ph.D.

THE YUCATAN HALL OF RECORDS:
CHICHEN ITZA'S TIME TRAVELER CULT

iUniverse books may be ordered through booksellers or by contacting:

iUniverse
1663 Liberty Drive
Bloomington, IN 47403
www.iuniverse.com
1-800-Authors (1-800-288-4677)

ISBN: 978-1-5320-0538-1 (sc)
ISBN: 978-1-5320-0539-8 (e)

Print information available on the last page.

iUniverse rev. date: 12/02/2016

ALICE AND AUGUSTUS LE PLONGEON (RIFLES NEARBY)
AT UXMAL, YUCATAN PENINSULA (1880'S).

This book is dedicated to Alice and Dr. Augustus Le Plongeon, who braved heat, fatigue, illness, bandits, and more, to unknowingly record some of the earliest observations of the Yucatan Hall of Records at Chichen Itza and Uxmal.

Special thanks to all of you… you know who you are, for listening to my ideas, and especially for debating with me.

"I don't see the logic of rejecting data just because they seem incredible."

Sir Fred Hoyle

GOLFO DE MEJICO
PLANO
DE
YUCATAN
1848
$N = R^{*} \cdot f_p \cdot n_e \cdot f_\ell \cdot f_i \cdot f_c \cdot L$
The Drake Equation:
N = the number of civilizations in our galaxy with which communication might be possible (i.e. which are on our current past light cone);
R_* = the average rate of star formation in our galaxy
f_p = the fraction of those stars that have planets
n_e = the average number of planets that can potentially support life per star that has planets
f_l = the fraction of planets that could support life that actually develop life at some point
f_i = the fraction of planets with life that actually go on to develop intelligent life (civilizations)
f_c = the fraction of civilizations that develop a technology that releases detectable signs of their existence into space
L = the length of time for which such civilizations release detectable signals into space

Contents

The Eastern Hemisphere

The North and South Poles

The Western Hemisphere: Europe

Africa

The Middle East: Israel and Sinai

The Atlantic Region

The Americas: The Southwestern U.S. and Old Mexico

The Americas: Central and South America

Introduction

Time Travel

In 2009, Holger Bech Nielsen—a father of string theory—and theoretical physicist Masao Ninomiya, suggested that CERN's Large Hadron Collider (LHC) could function as a time machine. They also hypothesized that humans in our future will send a bird backwards in time to sabotage the collider. Moreover, Stephen Hawking said time travel, "might be within our capabilities in the future." In addition, University of Connecticut physicist Ronald Mallett has been designing a time machine for over a decade. And, in 2010, news sources indicated that a man—claiming to be a time traveler—was arrested for trying to sabotage the LHC. Some argued that his effort was an April Fools prank, while others suggested the prank was a disguise for *actual* time travel. They cited the man's easy escape from CERN, a highly secure facility, as evidence.

Time travel is possible, according to Albert Einstein. And, while it seems unlikely that the CERN man or the bird were time travelers, the position in this discussion is that time travel is possible, and that it has already occurred. Consider the following theory.

In the future, a U.S. military officer will travel backwards in time. His evidence will be a photograph of a man next to a prehistoric kangaroo and a koala (and apple and orange trees), at a time before man existed on earth. The photo will be housed at the Smithsonian Museum. While the public may not ever see the photo, a drawing of the scene is already available. It is a century old, and will be shown later on in this book. The military man (and/or his crew) will also travel to more than 4,000 planets with intelligent life, and learn their histories and futures. For instance, one of the planets—replicated at the Mayan city of Coba—will be shown to have reached its classic period when earth was roughly in its Cretaceous period; before there were people here. The evidence is a stone map of earth the size of a bicycle tire. It shows Africa and Antarctica having mostly separated from the larger landmass. Meanwhile—judging by the images and steles—life on "Planet Coba" appears to have been thriving when earth was new. "Planet Coba's" crust was "thin" compared to earth's, and its tectonic plates had moved very little before a major earthquake—closer to its core than any on earth—destroyed much of its population. As will be discussed, the military officers instructed the Mayas to use ball court features to indicate the widths and depths of fault lines and earthquake zones. In the unusual case of Coba, one ball court has a stone skull—a Mayan symbol of death—embedded in its center.

The Mayan ball games are a combination of soccer and basketball, and are but one of a series of games whose rules are required to understand the time traveler's messages. The other games include, "Twenty Questions," "Trivial Pursuit," "Tic Tac Toe," jigsaw puzzle assembly, and more. Part of the challenge to the player is figuring out which game is being played at a given moment. The games have an apparent American bias, which will be explained shortly. They also incorporate math, geography, history, sociology, the arts, and other subjects typically found in a high school or college education. In other words, the time travelers—from earth in our future—chose to communicate with us through brain teasers based on sports and games.

However, to continue, the military time traveler's dilemma will be where and how to store the records of his travels and observations; including how to protect and preserve them. The solution will be the Mayas; bright, skillful people with knowledge upon which to build. Thus, roughly a thousand years ago (c.a. 900 AD), the time travelers persuaded the Mayas to record their planetary stories, one of which was partly presented a moment ago from Coba.

The military officer will introduce himself to the Mayas by the rank on his insignia "shield" (uniform patch). The Mayas will refer to him—and to other officers—as "Lord Shield," which, in their language is "Lord Pakal." The Mayan use of descriptive names for outsiders was presumably beneficial to the time travelers, as it would have minimized the risk of ancestral targeting. In other words, if time travel is achievable, a traveler might want to avoid revealing enough personal information to prevent his birth (i.e. because someone prevented his parents from meeting). Moreover, as will be shown later, Palenque's sarcophagus of Pakal is an encryption. It identifies him as an *American* who travels between worlds. Accordingly, while the military officer will describe himself as an "American," the Mayas will phonetically refer to his home in heaven as "Omeyocan." They will pronounce the word for "god," similarly. Thus, a "cargo cult" syndrome will ensue. Following is a brief description of the cargo cults.

During World War II, indigenous peoples in the South Pacific began seeing strange objects (military planes) in the sky. When the planes landed, the airmen shared their "cargo" (i.e. canned beans, etc.) with the indigenous people, who perceived them as their benevolent ancestors—gods, essentially. However, the end of the war brought the end of cargo, which the tribesmen sought to regain by intensifying their worship. So, they built life-size airplanes, landing strips, and control towers—out of bamboo—and incorporated them into their religion.

Accordingly, as the airmen were perceived as gods in the South Pacific, so will the military time travelers be perceived as gods to the Mayas. They will teach the Mayas the stories of other planets and the story of earth. However, their earth story will emphasize America—the United States. For instance, based on the time travelers' instructions, the Mayas will create a replica of modern day Washington, D.C. at Chichen Itza—a thousand years before Washington, D.C., itself, would be built. Similarly, the Mayan "creation" myths will allegorically refer to the creation of America—not the creation of the world. Thus, George Washington and John Adams will become the "Hero Twins" in a war against the "Lords of Xibalba"… the British House of Lords. England's King George III will be allegorically described as an arrogant "bird-demon" (who sits on a throne and has teeth) because King George III had chronic porphyria. At the time of the American Revolutionary War, porphyria was attributed to demon possession.

Similarly, Abraham Lincoln will become the "God of War" because of his role in the American Civil War. He will be memorialized at Chichen Itza's "Temple of the Bearded Man" at its Washington, D.C. replica—precisely where the Lincoln Memorial would stand—a thousand years later. Moreover, they sculpted a round object that, at certain times, projects an illuminated disk that encircles "The Bearded Man's" profile, thus creating the image of an American penny. In addition, the American Civil War will be fought between a "marshland capital" (Washington, D.C. was built among marshes) in the North, and a metaphoric land of African slaves in the South. As will be explained, the black slavery cue is the massive Olmec heads found in Veracruz. It was known in Pre-Columbian times—and in the civil war creation myth—as "Snake Hill," which is a reference to Snake Hill in Richmond, VA, the capital of the Confederacy. Thus, according to the myth, the civil war will ensue when the "Southerners" secede from the north and make plans to conquer Mexico City. In other words, the Mayan account is very similar to the American one, wherein the southern confederacy planned to conquer Mexico City as part of its "Golden Circle" expansion.

The American military officers will also instruct the Mayas to build a stonewall memorial to fallen soldiers (at Chichen Itza's Washington, D.C. replica), in the same spot where artist Maya Lin's Vietnam Memorial Wall would stand—a thousand years in the future. The officers will also instruct the Mayas to carve sixteen Secret Service counter-snipers—with modern combat equipment—on the roof of Chichen Itza's White House replica. They will symbolize the White House as a place where eagles live, because the presidential seal is an eagle. In addition, they will direct the Mayas to carve five tons of W-33 eight-inch (diameter) nuclear warhead replicas, and stack them inside the replica of the secret American nuclear arsenal in Iceland. The relevant geographic and architectural features can be observed in Washington, D.C. and Chichen Itza, and photos are presented in this book.

Essentially, to repeat, the U.S. military officers will build on the Mayas' excellent capabilities in construction, mathematics, language, writing, etc., and teach them more: earth's geography, extraterrestrial geography, history, future events, global cultural practices, etc. In turn, the Mayas will diligently record everything—in obedience to the "gods."

Over centuries, the Mayas will sacrifice human lives in the process of story-telling, which suggests that the American gods were malevolent. However, if their goal was to protect mankind from a worse evil (e.g. a plague, an extraterrestrial threat, etc.) in the distant future, then perhaps they were benevolent. In other words, as will be shown, most or all of the Mayan sacrifices were demonstrations of violent acts that either would, or did occur at specific locales, on specific planets. It was a type of record-keeping by pageantry. For instance, "sacrificial victims" were thrown into Chichen Itza's Sacred Cenote—but not into the other cenotes there—because the sacred one represents the circle of water around the North Pole—where the Inuit live. Their custom was to throw the elderly—who could no longer keep up with the group—into the icy waters, as a form of senicide. Similarly, evidence of child sacrifice at Chichen Itza has only been found at the South American and Sub-Saharan Africa replicas—where child sacrifice would be practiced in actuality in those places. Thus, for instance, the Mayas didn't randomly paint naked people blue and, simply put, chop their hearts out. Rather, they only did this at the temple that represents France (the "Temple of the Warriors"), which is historically accurate regarding France's Gaelic invaders. The Mayas learned all this from the time travelers.

Presumably, one of a time traveler's greatest threats would be his or her own mortality. Thus, since an American time traveler would be familiar with the Judeo-Christian creation myth about Adam and Eve, he would probably search for the "Tree of Life," and the "Tree of Knowledge," in the "Garden of Eden," even if only to disprove their existence. Apparently, however, the American military men will find the actual Garden of Eden; or, they will lead us to believe they did. They will construct a stone replica of the sacred garden in the Chichen Itza temple that represents Antarctica. The symbolism will be (is) difficult to miss: two strange-looking trees, an intelligent serpent (who walks upright and talks) who lives in them, a river that subsequently divides into four, at a place where people were said to have interacted with the divine—long ago. A photograph of the replica of the Tree of Life—known to the Mayas as the "Tree of the World"—and a photograph of the carved image of the "Vision Serpent," will be presented later on in this book. The garden temple is located deep inside Chichen Itza's Cave at Balankanche.

Finally, this discussion may not be the first one about the hypothesized time travelers. If the theory presented here is accurate, they could be the beings about which Sir Fred Hoyle, the Big Bang theorist, advised. According to journalist Jim Marrs, decades ago, Hoyle publically alerted us about the existence of space-time travelers. He said they have science and intelligence beyond our understanding, and that they have existed for a long time. Hoyle also said the beings influence world events, that we are "pawns" in their "great game," and that their existence is known to world governments.

Archaeological Discoveries

When the time travelers showed the Mayas how to create a world map at Chichen Itza, their Mediterranean Sea showed an island off the west coast of Sicily with a "Stonehenge" on it. Known today as "Pantelleria Vecchia Bank," the island sank 8,000 years before the Mayas mapped it. In 2015, archaeologists Emanuele Lodolo and Zvi Ben-Avraham reported finding a "Stonehenge" on the sunken island, and their research was published in the *Journal of Archaeological Science*. If the time traveler hypothesis is inaccurate, then how did the Mayas know about the island and its monoliths? Does clairvoyance offer such details?

Similarly, to elaborate on what was mentioned above, scientists in *Smithsonian Magazine* indicate that Australia's kangaroos and koalas came from South America millions of years ago, when the southern continents were fused (i.e. "Gondwana"). Plausibly, then, Mayas in the Yucatan Peninsula knew about them (somehow), which would explain their presence in a temple image at Chichen Itza. However, while no one knows when the last kangaroo and koala left Mexico—or if they even made it that far north—scientists imply that those particular marsupials died out in the Americas when the continents divided. That would have been about 60 million years ago—tens of millions of years before the first humans are believed to have appeared on earth (six million years ago). If that is accurate, then how did the Mayas—as recently as a thousand years ago—manage to draw a man next to images of a prehistoric kangaroo and koala in the upper portion of the House of the Jaguars?

There is more evidence. In a 2013 book about the Mayan map (*Time Traveler in the Age of Aquarius*), I hypothesized that Chichen Itza's Pyramid of Kukulkan is symbolically surrounded by water, and that its two snakes symbolically move across that body of water. My evidence was based solely on the Mayan map at Chichen Itza. Two years later, in 2015, archaeologist Dr. Andres Tejero-Andrade used sophisticated radar-type equipment to conclude that Kukulkan's pyramid had a water theme, which involved snakes moving in water. Our findings were not coincidental. There is a way to read and interpret the Mayan maps. However, it is a bit counterintuitive. For instance, while archaeologists indicate that Mayan ball court games represented warring tribes, it is proposed that the games actually represented "warring" tectonic plates; earthquakes reenacted or "pre-enacted" on massive planetary maps. However, before I explain about the Mayan maps in more detail, let us consider the New World priest, Fr. Diego de Landa.

Friar Diego de Landa

When Fr. de Landa learned that the 16th century Mayan Catholics were secretly worshipping their former gods, he burned their old religious books—twenty-seven of them, according to him—and he destroyed 5,000 of their statues. In Chichen Itza and across the Yucatan, the Mayan priests protested. And, that is how they tricked de Landa. For they had backup documents carved in stone, more than 4,000 of them, according to experts. Many of the documents had thousands of sections, and each section had many messages. The Mayas had foreseen de Landa's fiery Auto de fe ("Act of faith"), it is proposed, because they had been forewarned by the time travelers. So, centuries in advance, they prepared with Enochian redundancy; they put messages in both wood (books) and stone (temples)—and some of the books (e.g. the Dresden and Madrid codices) have survived. Thus, the American time travelers tricked the Mayas (into thinking they were gods) who recorded their stories, and the Mayas tricked the Spaniards, who sought to destroy those stories.

The Hall of Records: An Overview

This book interprets Chichen Itza's temples and features within the context of a new theory: that the Yucatan Peninsula is a (predominantly) Mayan "Hall of Records" in the New Age sense. It is a library of historical and prophetic accounts of earth and other planets. The theory is consistent with archaeologist Dr. Guillermo de Anda's proposition that the Mayas were trying to "represent the universe with their constructions."

Contemporary discussions about the Hall of Records say it contains answers to the mysteries of the universe, with an emphasis on earth. For instance, ancient astronaut theorist David Wilcock described the hall as a place containing "...all of the records that we would need to completely rebuild our history, and understand... widespread extraterrestrial presence here on earth." Since the main point of this discussion is the map and the Hall of Records, from this point on there will be decreasing discussion about the hypothesized time travelers.

The Three Halls of Records

According to legend and prophecy (e.g. Edgar Casey), long ago someone constructed three Halls of Records: one in Egypt, one in Bimini (the Bahamas), and one in the Yucatan. It isn't clear if the military men built all three halls, or only the one in the Yucatan.

The Egyptian and Bahamian Halls of Records

The Westcar Papyrus of the 4th Egyptian Dynasty and other sources (e.g. Edgar Casey), place the Egyptian Hall of Records beneath the Great Sphinx. In recent years archaeologists found a sarcophagus in the "Shaft of Osiris" 95 feet below the sphinx, but found no encyclopedic artifacts. However, the "Shadow of Osiris" that symbolically climbs up from the shaft and onto the Sphinx (Virgo and Leo)—on the summer solstice—suggests that the Egyptian Hall of Records is actually the zodiac. The zodiac, it has been argued, is the key to ancient mythology (e.g. Greek, Egyptian, etc.) which may actually be allegorical accounts (e.g. the Olympians and the Titans, etc.) of an even more distant ancient history. The rationale for connecting the Shaft of Osiris with the shadow that appears on the solstice is that the shadow was part of the Egyptian soul, the Sphinx was a trickster, and the light of Ra (sunlight) reveals truth.

The Shadow of Osiris (below) was photographed with a low-quality camera. It would have been clearer, had it been taken on a better camera.

In any case, it is most spectacular when viewed with the naked eye—in real time—from the Sphinx Guest House in Giza. The shadow pageant proceeds as follows. First, the torso of Osiris forms—followed by his head and legs. The legs and climbing posture only appear because of an enigmatic stone box on the Sphinx's side. The box is the strongest piece of evidence that the shadow features are deliberate, since it has no other known purpose. Presumably the box represents the sarcophagus of Osiris. However, to continue—the right arm of Osiris forms next, which then extends toward the Sphinx's neck. Finally, after about 20 minutes of "trying" to mount the Sphinx (to ride it to the heavens), and as if in despair, the head of Osiris tilts back. He looks up at the face of the Sphinx as if in a final plea to join the gods in the heavens. However, according to Egyptian mythology, Osiris was to rule the Underworld, not the heavens. Thus, his shadow gradually morphs into an indistinguishable mass—until next year's summer solstice.

If the zodiac theory presented above is correct, then two of the three halls have now been found. Researchers are still searching for the third hall on "Bimini Road," submerged in the Bahamas.

THE "SHADOW OF OSIRIS" CLIMBING FROM THE SHAFT OF OSIRIS, NECROPOLIS, GIZA PLATEAU, SUNRISE ON THE SUMMER SOLSTICE (2010). PHOTO BY AUTHOR.

The Mayan Map: Micro

Contrary to legend and popular belief, the Yucatan Hall of Records is not a room filled with ancient tablets and scrolls. Rather, it is a walkable, schematic cartogram—a geo-social map comprised of massive, stone temple cities. The cities (maps) of Chichen Itza and Uxmal represent earth. The key to reading them is knowledge of world history and geography. For instance, as Linda Schele and Peter Matthews (1998) suggested, the Mayas employed "pyramid-mountains and plaza-seas" to tell their stories. In other words, Schele and Matthews understood that the Mayas were using temple structures to present landscapes, which are the foundation for maps. Accordingly, Chichen Itza's *Pyramid* of Kukulkan represents Mt. Pico in the Azores, and its surrounding "Grand *Plaza*" represents the Atlantic Ocean.

The Mayan Map: Macro

Chichen Itza and Uxmal are the keys to, and the components of, the larger universal map. This larger map is comprised of all Mayan temple cities. The ones beyond Chichen Itza and Uxmal represent other planets. Thus, the images, hieroglyphics, etc. in those other temple cities describe extraterrestrial life. In other words, a thousand years ago the Mayas didn't simply create a map of the world. They created a map of the universe, and they placed earth (Chichen Itza and Uxmal) within it. The earth map is the microcosmic legend for interpreting the macrocosmic universal map. Thus, pyramids can represent mountains on any planet. Similarly, since ball courts at Chichen Itza and Uxmal represent tectonic plates and earthquake zones on earth, they presumably represent the same phenomena on other planetary replicas.

Concealing the Mayan Map

As will be shown, the maps at Chichen Itza and Uxmal match our contemporary maps, but they are difficult for Westerners to identify, partly because of their artistic style. Thus, the map—the heart of Mayan prophecy—had been at de Landa's fingertips all along, but he hadn't noticed it because he had no concept of a global map the size of a city, or a country represented by a large temple. European maps were made of paper or fabric and displayed on tabletops. They were not massive temples a square mile and made of stone. Moreover, they were not made by ostensibly primitive people. Similarly, the Mayas' "*Animal Farm*" allegories (anchored in the Chichen Itza map as *myths* and *rituals)* would have meant nothing to de Landa. That the Mayas named one of their major cities "What Will Come—the Future," would have been xenophobically dismissed as more evidence of Mayan quirkiness, rather than as a prophetic account of China, Angkor, India, Thailand—Asia.

Apart from the map, the Mayan myths are without any moral or ethical takeaways, which supports the idea that the stories are based on real events. For instance, in one of the central myths, 400 men plot to murder a man who had just done them a large favor. Alone, the story has no apparent meaning. However, in light of American history its allegorical meaning is clear. It is about a specific group of 400 men who played a major role in American history. They will be discussed later on. The point, now, is that the Mayan *myths* describe actual events in real places (e.g. nations) that are portrayed on the map. Similarly, the ritualistic *sacrifices* at Chichen Itza and Uxmal were not efforts to appease angry gods. Rather, they were pageants about deadly events (i.e. wars, earthquakes, senicide, etc.) in earth's past and future. Thus, as mentioned previously, it is not coincidental that the Mayas disrobed their enemies (and, volunteers, according to de Landa), painted them blue, and killed ("sacrificed") them at the Temple of the Warriors. The rationale is that this particular temple represents France—where Gaelic invaders historically disrobed, painted themselves blue, and attacked France in the nude.

The Mayan *map*, *myths*, and *rituals* are parts of a whole (the Hall of Records), and no part can be understood without the others. Thus, the three represent a media encryption strategy to protect the messages in the Hall of Records (e.g. from de Landa). The hall was hidden in plain sight long before the Spanish conquistadors and de Landa arrived, and nature helped protect it with apocalyptic seals made of jungle overgrowth.

Allegorical encryptions are not uncommon in cults, religion, etc. Jesus said he spoke in parables because some of his messages were for select audiences, only. Similarly, since animals with x number of heads, y number of horns, and z crowns—as in St. John's apocalypse—probably don't exist, such images are probably symbolic. Another reason for communicating in symbols is that humans apparently think (e.g. the "Freudian slip") and dream in symbols. Thus, while there is no apparent necessity for animals, plants, keys, etc. to symbolize the U.S. President, political parties, the Vatican, sports teams, automobiles, etc., the practice is common.

Predictive Maps

In this discussion, a predictive (or "prophetic") map is one that includes features—whether natural or man-made—that did not yet exist when the map was created. For instance, the Chichen Itza map includes a replica of the English Channel Tunnel, which wasn't constructed until a thousand years after Chichen Itza was built.

Eastern and Western Hemispheres

Chichen Itza is a predictive map of the Western Hemisphere. Its original name, "Seven Great House," may have referred to Chichen Itza's future status as one of the Seven Wonders of the World. Similarly—as stated above—the Mayan temple city of Uxmal (meaning, "What will come, the future") is a predictive map of the Eastern Hemisphere.

Chichen Itza's map of the Western Hemisphere begins at a replica of the Adriatic Sea ("Unnamed Ball Court") and runs to a replica of California ("House of the Deer"). The Eastern Hemisphere at Uxmal begins at the Russian replica ("The North Group"), and runs to Thailand ("Temple of the Phalli"). It includes China ("House of the Fortuneteller/ Magician") and more. Together, the two hemispheric maps form the world map. In turn, the world map forms the legend to the larger, interplanetary map. This interplanetary map consists of the Yucatan Peninsula's roughly 4,000 (4,400 according to Witschey and Brown) temple cities, including Chichen Itza and Uxmal. Thus, as Giordano Bruno suggested long ago, there is a lot of intelligent life in the universe.

Yucatan Nations

The nations comprising the Yucatan Peninsula (the Hall of Records) on our contemporary maps include: Mexico, Guatemala, Honduras, Belize, and El Salvador.

Aztecs, Toltecs and Others

The Hall of Records may not be limited to the Yucatan Peninsula. For instance, Mexico's northern temple cities seem to contain important messages as well. These cities include Teotihuacan, Tula, and others in the general region of Mexico City. They are not included in this discussion because they may not represent planets. In general, the Mayan cities seem to represent planets, while the non-Mayan cities seem to represent concepts and processes. There are exceptions, however. For instance, Tula, just outside of Mexico City, is home of the Toltec colossi. It was apparently built by the Toltecs, rather than by the Mayas, and it offers ball courts, which indicate planetary surfaces.

TOLTEC WARRIORS, TULA, MEXICO,
PHOTO BY AUTHOR (2015).

Teotihuacan

Teotihuacan was once the home of a phallic cult whose orientation is apparent in its sexually explicit art. Similarly, the layout of the city models the human reproduction process. It is difficult to observe because the organ replicas are squared rather than rounded, and some of the telltale structures have been removed. For instance, the moon goddess who once stood at the end of the "Avenue of Death" (a 1 kilometer journey consisting of exhausting obstacles—especially at high altitude) is now in a museum in Mexico City. Also, most discussions and photographs of Teotihuacan focus on the avenue, and ignore the two temples at its beginning—the testicle replicas.

Without presenting too many details here, visitors who walk the Avenue of Death are inadvertent participants in a human biology pageant—as male seeds. Consider, for instance, Karl Taube's discussion of one of the symbolic Teotihuacan murals. It shows "fish-men" "swimming" through an "opening" to another world. In other words, "Avenue of Death" refers to the fate of the seeds that don't fertilize the egg. Thus, apparently Teotihuacan's pyramid complex was once a competitive obstacle course, where the winner was the first contestant to reach the square egg-like object at the Earth platform. Stepping onto it, therefore, would have represented birth. Accordingly, "Teotihuacan" means "birthplace of the gods," which could mean people are "gods" in some sense. The point here, however, is that Teotihuacan does not appear to represent a planet as do the Mayan cities. Thus, as Teotihuacan is not Mayan, it raises the possibility that the other non-Mayan cities are not planetary replicas either.

The Olmecs

The two prevailing schools of thought regarding the Olmecs are Dr. Van Sertima's theory that ancient Africans sailed to the Americas, and, simply put, the opposing view is that they did not. At the center of the discussion are massive stone heads that look African. Thus, the logic is that the Olmecs carved images of themselves, which were subsequently discovered centuries later.

In light of the map theory, however, an alternative explanation is offered: the Olmec statues were images of the African slaves who were yet to arrive in Mexico—in the future, rather than images of Africans who had lived there in the past. Consider the evidence. The Olmec images were found in the Veracruz region, where African slaves were brought, including those who were subsequently moved to other parts of the Yucatan. Moreover, unlike other statues in the Yucatan Peninsula, the Olmec ones consistently look strong, and typically unhappy—like slaves. The statues are never in temple cities, which means they symbolically have no homes. They have no bodies (just heads), which means they are immobile—they cannot go anywhere—like slaves. Similarly, many of the African slaves in Mexico worked in mines, which explains the helmets on many of the Olmec statues. However, since not all slaves worked in mining, not all stone heads are wearing helmets. One statue, "El Negro," wears no helmet, and his arms are bound like a slave's. His image supports the slave-prediction hypothesis more than any other, as it is unclear why the Olmecs would honor themselves by creating such an image. It will be discussed later on in the American Civil War section. Finally, consider that, according to archaeologists, some or all of the Olmec images were deliberately buried after they were carved. Stone heads weighing many tons were placed on circular structures several meters below ground, and then covered up. The stone heads were found a century after the African slaves arrived in Mexico. Therefore, it is impossible that the slaves carved them. Thus, evidence that ancient Africans, or more recent African slaves carved the stone heads, is weakened.

Weakness of the Map Theory

The map theory's weakness is that it is based on the historically questionable concept of a Hall of Records. A few scientists searched for the hall in Piedras Negras, Guatemala, or otherwise tried to prove its existence. However, historically it has been regarded as mythological in most discussions that demand hard evidence.

Implications of the Map Theory

At a mundane level, the map theory has implications for politics, international relations and business. For instance, what if France had known about the Nazi invasion before Hitler planned it? Would Cold War Russia have attacked Iceland had they known of America's nuclear arsenal there? Would England have fought the American Revolutionary War, had they known about Lord Cornwallis' defeat before the war began? What real estate investments would have ensued, had speculators known about the English Channel Tunnel in advance? Does the tiny Trinity Test Site (Alamogordo, NM) replica mean we were warned about nuclear weapons? Is it coincidental that the Alamogordo test site was located in a desert called "A Single Day's Journey to Death," given that its Mayan replica was called a "funerary" structure?

Then there is the larger question. If the map theory is valid, and there is intelligent extraterrestrial life, then why hasn't anyone contacted us? One possible answer can be found in the research of Guglielmo Marconi and Nicola Tesla. Marconi won a Nobel Prize for his work in radio communication, and Tesla was a genius in radio and electronics as well. Each, independently and years apart, said they received extraterrestrial radio signals. Tesla reported receiving the message "1, 2, 3." However, society essentially ignored their claims.

The Origin of this Book

The map theory was developed inductively while researching Chichen Itza for my "Davenport Prophecy" novel series. Personally, I was searching for the backstory to the Book of Genesis—where a man and his wife are happy in a garden; suddenly there is a talking snake, and then everything goes bad. I wanted to know more, as that basic story was not only in the Bible, but in other ancient sources as well. In any case, I interpreted the initial similarities between Chichen Itza and the world map as coincidental. However, after a year of steady research, I kept finding coincidences. Eventually there were simply too many of them.

I originally published the map theory as a nonfiction application of the Torah Code theory. Each theory was conceptually defined within the Akashic Records paradigm—purportedly the governing software behind reality. *Time Traveler in the Age of Aquarius* theorizes that the Hebrew Resh expands the Torah Code matrix into the Akashic Records (without which time travel would be riskier). However, I no longer agree with some of my previous interpretations. Accordingly, this new book presents a simpler, and hopefully more accurate interpretation of the Mayan map. It focuses on Chichen Itza, a map of earth's Western Hemisphere, but it includes a brief discussion about the Eastern Hemisphere's map at Uxmal.

I found the two hemispheric maps by studying the Le Plongeon's research. They knew they were on to something. However, they couldn't put their finger on it, because the structures that would have indicated time anomalies had not yet been built. For instance, the Le Plongeons could not have known that Chichen Itza's Tzompantli (wall of dead soldiers) represented the Vietnam Memorial Wall, because the war had not yet been fought, let alone the construction of its memorial.

Neither could they have surmised that the Chac Mool 25 feet under the Great Ball Court represented the Washington, D.C. Cold War fallout shelter—because there were no Cold War nuclear bomb shelters in their day. The Le Plongeons lived and died before the world's first nuclear explosion at Alamogordo, NM, which is why the replica of the Alamogordo bomb bunker (at Chichen Itza) didn't register with them; nor did the five tons of nuclear bomb replicas that they found at the Iceland replica.

Still, however, the Le Plongeons are the unsung heroes of archaeology. Theirs' is a romantic tale of high intellectual adventure, discovery, and loss. Augustus was an English medical doctor who had already proved his mettle in rugged situations, and Alice was an English photographer 25 years younger than he. They met when she was 19, and he basically invited her to help him uncover the mysteries of the universe—by studying ancient sites—and she could be their photographer. She agreed, and off they went. They traveled, dug, photographed, cajoled governments for permission, and published. At first the world liked their work. However, when the Le Plongeons started subjectively interpreting their findings (e.g. Jesus' lament on the cross was in "pure Maya" rather than Aramaic or Hebrew), the scientific community began withdrawing.

After Augustus died, and Alice was elderly in her Brooklyn, NY apartment, she gave their Yucatan photographs—their life's work—to her best friend, a young woman named Maude Blackwell. She told Maude that if the world ever became interested in the Mayas again, then she should make the images available. Otherwise, she should burn them.

Fortunately, the world did take renewed interest in the Mayas. Perhaps for some it began with the violent old film, *Kings of the Sun*, or with the romantic thriller *Against All Odds* a few decades later. Perhaps others heard reports of a serpent shadow on the equinox in the ancient part of Mexico, while others may have happened upon excavation photos online, compliments of the Getty Library, the Cornell University Library, and others.

If the map theory is correct, then the pioneering scientists of extraterrestrial life may not be astronomers and astrophysicists, but rather archaeologists and anthropologists. Moreover, it could mean the architect (who *may* be the hypothesized military officer) of the Mayan temples has shown us the player's manual to Hoyle's "great game," wherein we humans are "mere pawns" in the library of Zardoz; and the finger beckons. There is no Faustian bargain to be made here—nothing that can't be found with the internet, a tourist visa, and some occasional bushwhacking. The ancient Mayan library is as accessible as any other (except, currently some places are more dangerous than others. Chichen Itza is not on the list of dangerous places.)

The Chichen Itza map, below, is a very small two-dimensional map of a very large three-dimensional map of the Western Hemisphere. Thus, it is a map of a map. However, it includes a few necessary distortions so it can fit on a page.

Since many of Chichen Itza's structures and artifacts have been lost, destroyed or removed to museums, the effect has been that of reading a biography from which pages have been randomly torn. For instance, a tour of the "Platform of Venus" (Iceland's replica) should reveal—as mentioned previously—five tons of *red* and *blue* stone replicas of 8-inch diameter nuclear warheads… against a *white* background. It should also reveal a dozen flying vipers in the same colors as U.S. Air Force pilots, a stone lion wearing a sun hat, a sculpture of a contorted infant boy with realistic eyes and finger nails, and a reddish stone slab with a fish on it. However, none of those artifacts is still at the platform. With the exception of a few remaining warhead replicas nearby, they have all been removed. Their images are included in my map because they tell the story of the Western Hemisphere. Unfortunately, there are many other temples at Chichen Itza from which "pages" have been torn. Hence, the importance of the Le Plongeon's photos and field notes. They help us identify the map's missing pieces, which—according to the map theory—enable us to understand earth's past and future. The Le Plongeon's research is the key to understanding Chichen Itza and Uxmal. In turn, those temple complexes are the keys to interpreting the stories of other planets in the Yucatan Hall of Records.

There is clearly something "funny" going on, and the Yucatan Peninsula is in the middle of it. It is either a map of some portion of the universe, or someone wanted us to think it is. There are too many similarities between the Mayan structures and world geography and history for it to be otherwise. There are too many coincidences.

To me, the most fascinating aspect of the map is its inclusion of a variety of sacred places from Buddhism, Judaism, Christianity, Hinduism, Native American culture, the arctic home of the Inuit Elders, and more. Whoever created the map knew what we would believe, and where our holy places would be. He or she also knew where we would fight our wars, and who would win. The frightening question is—what did he or she know, versus what did he or she cause? Was there a Butterfly Effect?

§

Note: the best way to get a sense of the scale of the Mayan map of our Western Hemisphere is to use the map on the following page while touring Chichen Itza. Alternatively, one can virtually tour many parts of Chichen Itza—at ground level—in Google Maps. Unfortunately, many of the statues, steles, etc. on the following map are now in museums.

§

"We are all agreed that your theory is crazy. The question which divides us is whether it is crazy enough to have a chance of being correct."

Niels Bohr

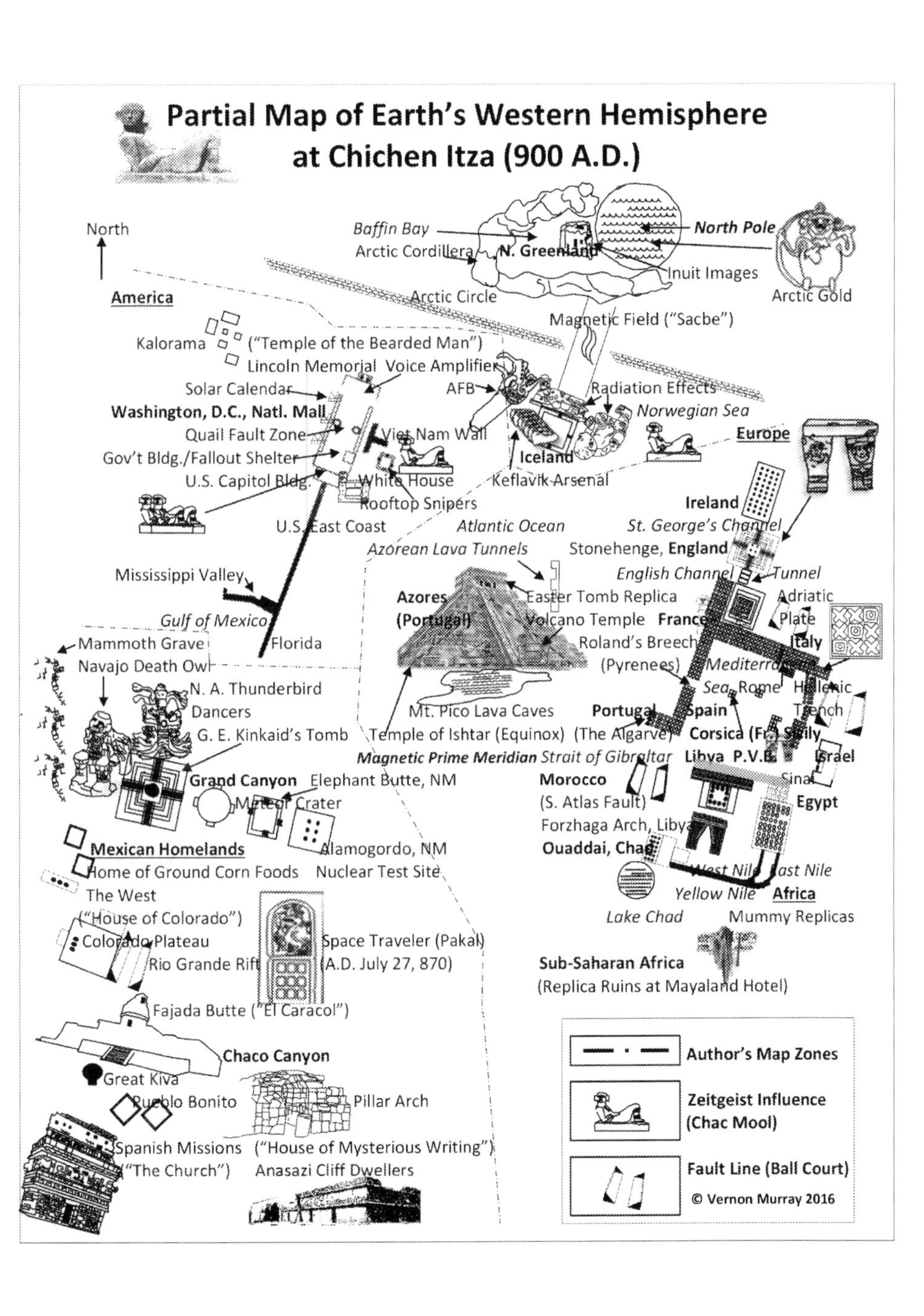

Partial Map of Earth's Western Hemisphere
at Chichen Itza (900 A.D.)
North
Baffin Bay
Arctic Cordillera
N. Greenland
North Pole
Inuit Images
Arctic Gold
America
Arctic Circle
Magnetic Field ("Sacbe")
Kalorama
("Temple of the Bearded Man")
Lincoln Memorial
Voice Amplifier
Solar Calendar
AFB
Radiation Effects
Washington, D.C., Natl. Mall
Norwegian Sea
Quail Fault Zone
Viet Nam Wall
Europe
Gov't Bldg./Fallout Shelter
Iceland
U.S. Capitol Bldg.
White House
Keflavik Arsenal
Rooftop Snipers
Ireland
U.S. East Coast
Atlantic Ocean
St. George's Channel
Azorean Lava Tunnels
Stonehenge, England
Mississippi Valley
English Channel
Tunnel
Azores
(Portugal)
Easter Tomb Replica
Adriatic
Gulf of Mexico
Volcano Temple
France
Plate
Mammoth Grave
Florida
Roland's Breech
Italy
Navajo Death Owl
(Pyrenees)
Mediterranean
N. A. Thunderbird
Sea
Rome
Hellenic
Dancers
Mt. Pico Lava Caves
Portugal
Spain
Trench
G. E. Kinkaid's Tomb
Temple of Ishtar (Equinox)
(The Algarve)
Corsica (Fr.)
Sicily
Magnetic Prime Meridian
Strait of Gibraltar
Libya
P.V.B.
Israel
Grand Canyon
Elephant Butte, NM
Morocco
Sinai
Meteor Crater
(S. Atlas Fault)
Egypt
Forzhaga Arch, Libya
Mexican Homelands
Alamogordo, NM
Ouaddai, Chad
Home of Ground Corn Foods
Nuclear Test Site
West Nile
East Nile
The West
Yellow Nile
Africa
("House of Colorado")
Lake Chad
Mummy Replicas
Colorado Plateau
Space Traveler (Pakal)
Rio Grande Rift
(A.D. July 27, 870)
Sub-Saharan Africa
(Replica Ruins at Mayaland Hotel)
Fajada Butte ("El Caracol")
Chaco Canyon
Author's Map Zones
Great Kiva
Zeitgeist Influence
(Chac Mool)
Pueblo Bonito
Pillar Arch
Spanish Missions
("House of Mysterious Writing")
("The Church")
Anasazi Cliff Dwellers
Fault Line (Ball Court)
© Vernon Murray 2016

Chapter 1

How to Read the Mayan Map

Overview

The Mayan map is read like most other maps, except it is part of a whole—the Yucatan Hall of Records. The hall consists of the map, rituals, myths, the five elements (earth, wind, fire, water and sky—"the Akash"), and time—the fourth dimension. The rituals and myths cannot be understood apart from the map, and the map cannot be read apart from the five dimensions and time. The time component means, for instance, no one could have identified the White House replica prior to the construction of the actual White House in 1792. Neither could they have identified the Secret Service rooftop snipers at the White House replica prior to 1901 when the Secret Service began.

An important difference between the Chichen Itza map and our modern maps is that the Chichen Itza map shows the earth's curvature. The perceptual impact is that some of Chichen Itza's objects are more southwestern than they would be on our maps. For instance, below, Arizona is *southwest* of Washington, D.C. on the modern map, but it is *south-southwest* on the Mayan map. Hence, the curved line.

Mayan vs. Modern Maps

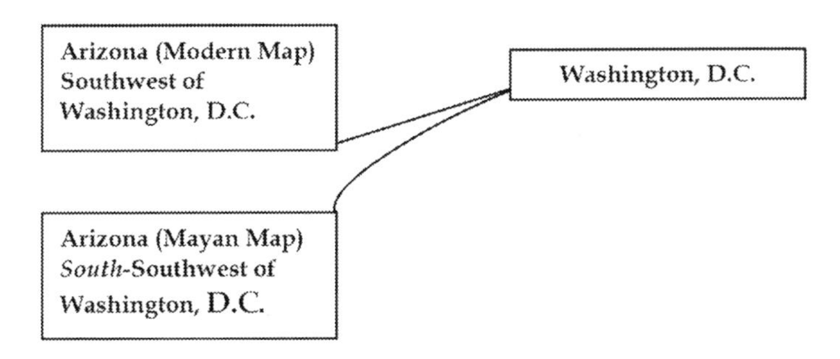

Reading the Mayan Map's Colors and Lakes

According to archaeologists, Chichen Itza's Sacred Lake of the north corresponds with the color white. The temple complex's other lakes correspond with red, yellow and black, for east, south and west, respectively. The central lake—beneath the Pyramid of Kukulkan—corresponds with blue-green—the center of a diamond-shape made of lakes. The ancient Mayas called the lakes "cenotes." Thus, the cenotes, colors and directions at Chichen Itza all correspond with locations and climates on the map of the Western Hemisphere. In other words, the map's legend at Chichen Itza has an even more abstract legend, which is comprised of a color-coded diamond from Mayan antiquity. The connections between the abstract Mayan map and the landscape of our Western Hemisphere are presented below.

Reading the Mayan Map's Color Code

With the exception of "red-east rising sun," each color-direction (presented above) is reinterpreted below. Accordingly, *white* represents the frozen north (e.g. Iceland). *Yellow* represents the warm equatorial sun to the south. East is *red* for sunrise. West is *black* for night (sunset), and the *blue-green* center represents the Atlantic Ocean at the Azores. Thus, the Pyramid of Kukulkan—at the center of the diamond (built over the 5th cenote at Chichen Itza)—represents the central orientation point of the Western Hemisphere—the Azores. The architect (via Mayan mythology) fixed each directional point with a specific color to cryptically help us identify Chichen Itza as a map. In other words, the map is to be anchored in the cenotes. Let us now briefly discuss the relationships among the Mayan map, myths, rituals, and time.

The Mayan "Diamond" Map of Directions, Colors and Cenotes

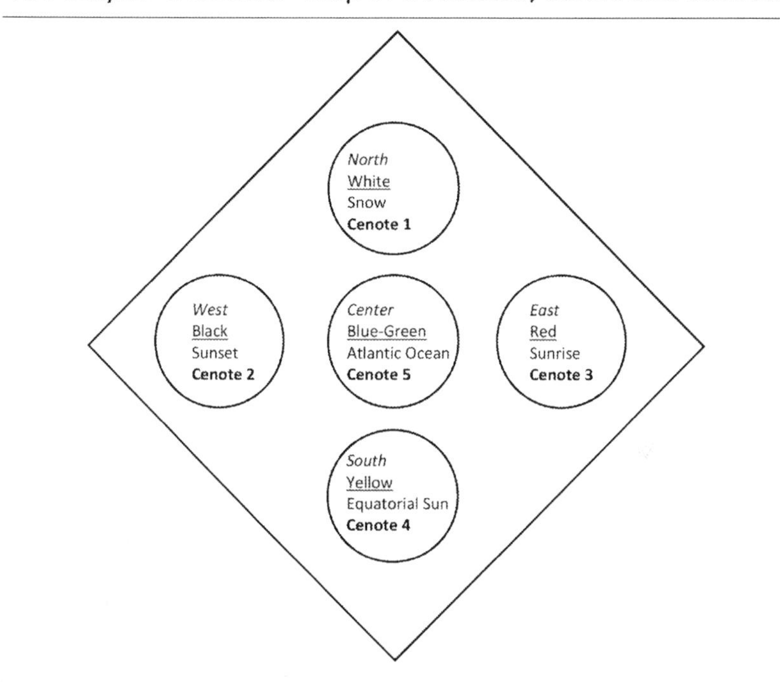

Note: Interpretations (e.g. "Atlantic Ocean") are the author's.

Reading the Mayan Map's "White Paths"

The Azores is centralized on the Mayan map of the Western Hemisphere because it is the natural Prime Meridian. "Natural" means it offers the most stable magnetic compass readings at sea. Thus, the Azores had been home to the Prime Meridian until international politics moved it to Greenwich, England in the 1800's.

The Prime Meridian's stability in the Azores comes from the magnetic North Pole's superior pull in the Mid-Atlantic region. It is why compass needles point north instead of in several directions, even though earth's magnetic fields run in several directions. Accordingly, the Mayan map at Chichen Itza employed eighty "White Paths" (called "Sacbes") that run in multiple directions to represent earth's magnetic fields. The largest white path aligns the Pyramid of Kukulkan (the

Azores replica) with the "Sacred Lake" (the North Pole replica), to indicate the strong magnetic pull of the north.

The Prime Meridian is marked on the Mayan map because our maps of the Western Hemisphere mark the Prime Meridian. In the first set of maps presented as a modern atlas, Abraham Ortelius placed the Prime Meridian (at 360⁰ on the map below) between the islands of São Miguel and Santa Maria in the Azores. The purpose of the sacbes may not have been solely cartographic, however. Since planetary magnetic fields are required to support intelligent life, perhaps the Mayan map included sacbes to help us identify inhabitable planets.

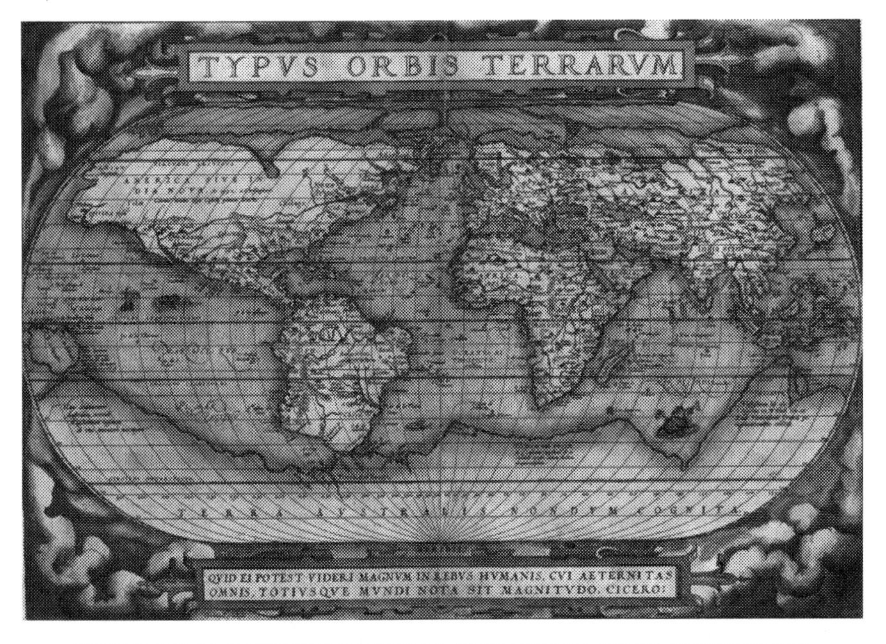

THEATRUM ORBIS TERRARUM ("THEATRE OF THE WORLD")
BY ABRAHAM ORTELIUS, SHOWS THE AZORES
AS THE PRIME MERIDIAN (360⁰), 1570.

The Map-Ritual Connection

Most or all of Chichen Itza's temple-specific rituals and sacrifices were "pre-enactments" and reenactments of past and future world events. For example, when the jaguar priests disrobed their enemies and painted them blue before extracting their (still beating) hearts, they were presenting a pageant about French-Gaelic military history. Long ago, Gaelic invaders would remove their clothing, paint themselves in (woad) blue, and—in the nude—attack France. Hence, this particular Mayan ritual was only performed at the "Temple of the Warriors" which represents France. The heart removal aspect of the ritual refers to the Gaelic hero, Robert the Bruce. His will stipulated that upon his death, his heart was to be extracted and carried into future battles.

Similarly, the Mayan ball courts resemble fault lines (and were called "crevices" in the Maya language), because they represent earthquake zones. Their deadly ball games were re-enactments and pre-enactments of earthquakes in such specific locations as Crete, Qinghai-China, and Morocco. The games were deadly pageants because earthquakes are deadly. If that hypothesis is right, then there will be (or once was) a major earthquake along the Rio Grande Rift along the Colorado Plateau—in the United States.

It is proposed, therefore, that some ancient Mayas lived dual lives as both regular people and as pageant actors. If that is true, then behavior in a given social context (e.g. an intertribal fight) could have been personal or "scripted." Thus, depending on the script, some "actors" were literally killed in their roles. However, they were not actors in the modern, Western entertainment sense. Rather, as the pageants and rituals were presumably religious, the actors were religious participants whose priests required their cooperation. Some victims, therefore, were essentially martyrs for the Hall of Records.

The Map-Myth Connection

Some or all of the Mayan myths are allegorical accounts of past or future events. For example, the strange story about the "Four Hundred Drunken Young Men" is actually about the "Maryland 400" of the American Revolutionary War. Consider the evidence. Since people cannot become star systems, the myth's account of the Maya 400 becoming the Pleiades is presumably allegorical. The Pleiades represent the point around which the Mayan sky rotates. Similarly, as will be explained later, the Maryland 400 played a pivotal role in saving the

American army from the British. Accordingly, the Mayan 400's ultimate foe—an arrogant bird who (unlike normal birds) had teeth, sat on a throne, had a private physician, and was demon-possessed, represented King George III (England's king during the American Revolutionary War), whose physician diagnosed his porphyria (disease) as demon-possession. His doctor used to beat him brutally—as was the common practice back then—in an attempt to "beat the devil out of him."

The above interpretations are plausible only because they are anchored in myths about the "Great Ball Court," which represents Washington, D.C. Temple-specific myths are an important tool for understanding the Hall of Records (i.e. the combination of temples, myths, rituals, etc.). As will be shown, some odd Mayan phenomena can be easily understood in light of the map.

The Map and the Five Elements

Chichen Itza's seemingly random images of "Chaac" the rain god are actually instructions to observe those parts of the map during rain storms—when flooding (*water*) reveals its seas and oceans. That is when, for instance, the Pyramid of Kukulkan is transformed into the Azores—in the middle of the Atlantic Ocean—as the "Grand Plaza" floods. Heavy rains also create an aqueduct in "Rome," and "Antarctica" emerges as an island at Chichen Itza's southern end (at Balankanche). Similarly, Chaac's images at the "Market" (the Libyan coast) and at other places on the map have been cryptically interpreting the flooding process for over a thousand years.

Earth provides most of the map's information in the form stone temples. Next, Kukulkan's equinoctial trek in light and shadow requires *fire* (the sun), which is the third element. Since Kukulkan's "voice" is heard when one claps hands at his pyramid, *wind*, which carries sound, is the fourth element. Without fire and wind, we would not be able to observe the annual reenactments of Ishtar's descent, and Christ's resurrection celebration, at the Azores replica. The fifth element required to observe the Mayan map is the sky—the "*Akash*." High-elevation imaging (e.g. from satellites) enables us to identify features on the Mayan map. The role of time in understanding the map is discussed in the next chapter.

Chapter 2

Time Travel

Forwards vs. Backwards Time Travel

This book offers evidence that someone from our future traveled to our distant past to leave information for us—the Yucatan Hall of Records. While many quantum physicists doubt the possibility of backward time travel, Dr. Ronald Mallett at the University of Connecticut is an exception. He believes time is a 4th dimension, which our three-dimensional existence prevents us from fully perceiving. Analogously, a hypothetical two-dimensional being cannot see us, because he cannot see up or down—only across. He can only perceive a line, and, in theory, lines have no width, no depth. Professor Mallett's theory supports backward time travel, so much so that he is currently designing a time machine. He has been published in academic journals, and interviewed by Bloomberg, the Washington Post, the Huffington Post, etc. Finally, consider that if time travel does not currently exist, then it never has and never will, anywhere in the universe, even though it is theoretically possible.

Time Travel Anecdotes

Native Americans in the Southwestern U.S. said their ancestors had been instructed by men from the future known as "Wing-makers." Similarly, Air Force soldier James Penniston suggested that his UFO encounter in Rendlesham Forest had been with a time traveler from earth's future—not with extraterrestrials. Consider also the images of modern, military-style, boats and aircraft from ancient Abydos. And again, apparently Fred Hoyle believed time travelers had been influencing mankind for a long time.

The FTL Anomaly

In 2011, scientists at the Italian National Laboratory reported receiving atomic particles from Geneva's CERN laboratory at a rate faster-than-light (FTL) speed, which is a requirement for time travel. However, a few months later the scientists changed their story, and attributed their initial findings to faulty optic wiring. The important question is—if a scientist were to publically announce successful time travel, would his or her government subsequently rescind the message? If time travel isn't the ultimate weapon, then it is not far from it.

Time Travel vs Clairvoyance

The map theory incorporates backward time travel as the default alternative to clairvoyance and coincidence. The map—it is argued—offers details about people (e.g. Napoleon's Corsican origin), artifacts (e.g. twin leopards on the east side of the palace of the Dalai Lama in Lhasa, Tibet), events (e.g. Germany's World War II invasion of France), and places (e.g. Portugal's Algarve) that exceed the presentations of any known clairvoyants.

As America's leading clairvoyant, Edgar Cayce accurately described the Hall of Records as being "without walls or ceiling" (there are no walls around the Yucatan Peninsula) and in the vicinity of Piedras Negras (in the middle of the Yucatan). However, consider that as accurate as Edgar Casey may have been, the map offers tangible evidence in stone. Now consider Nostradamus. He was a notable clairvoyant, but there are ambiguities in his predictions. Similarly, Jung's Collective Unconscious (the Akashic Records) is an unlikely source of the map because, presumably, it is the same as clairvoyance. Likewise, an excessive number of chance similarities between the Mayan map and our contemporary maps is ruled out as improbable.

Time Travel: the Pressing Questions

The idea of active, influential, time travelers can be frightening. Can they change history? Can they create false data, or masquerade as deities? Have they? Who were the Biblical "Sons of God," who married Earth women? Could a "Butterfly Effect" change history? The short answer to the above questions is: possibly, yes.

Chapter 3

The Architect and Extraterrestrial Life

The Architect

The designer of the Yucatan Hall of Records is referred to as "the architect" because he or she largely communicated via structures. In other words, while the hypothesized military time traveler is the Mayan contact, he is not necessarily the architect. The architect could be a woman. However, to continue, according to Maya Revival architect Robert Stacy-Judd, "architecture is a language," wherein "measures and motifs are words and sentences." Over a century ago, Yucatan archaeologist Augustus Le Plongeon described architecture as "more easily understood than language." Searching for answers to the great questions, he referred to the Yucatan as "a rich living current of occult wisdom and practice, with its sources in an extremely ancient past, far beyond the purview of ordinary historical research." Similarly, Chumayel (a Maya priest), said the architecture at Mayapan told the story of "El Dios," the Christian deity.

The architect also communicated with words and pictures. The late Phillip Coppens—an ancient astronaut theorist—suggested that a portion of ancient Maya writings indicated that "the gods" used Chichen Itza as a venue to tell mankind our future. Within the context of prediction, consider the Hall of Records' account of the World War II German invasion of France. It described people running and screaming and carrying all their possessions, attacking soldiers from the north pursuing and capturing members of one specific ethnic group (leaving everyone else alone), and sailors from the west crossing an ocean to enter the war zone. The related images showed mixed-race families fleeing with their children, and men with helmets (domed hats) and backpacks carrying sticks that produce smoke at one end—rifles—on a beach. One image shows a man wearing a hat similar to those of World War I England—army surplus. The mysterious architect was clearly knowledgeable about world events. Less clear was his motive for creating the Hall of Records, whether or not he distorted any information, and whether he was friend or foe to mankind. There is no telling, as his or her intellectual abilities may be far beyond ours.'

Extraterrestrial Life

While ancient replicas of the English Channel Tunnel, the Lincoln Memorial, and Napoleon's statue at Corsica are interesting, the Chichen Itza and Uxmal maps are merely components and legends of a larger map. The larger map consists of roughly 4,000 Yucatan cities (based on Witschey and Brown's estimate) which—it is proposed—are maps of other planets with intelligent life. Their stories are told in the images, hieroglyphics, and accompanying myths and rituals in the Yucatan. Thus, the Yucatan map is like an ancient university library—where one studied the universe. As archaeologist Dr. Guillermo de Anda suggested, the Mayas were trying to "represent the universe with their constructions."

Mars

Palenque's "Lord Pakal" and "Red Queen" were buried in bright red cinnabar to indicate that they were buried on Mars—the "Red Planet." The two small naturally-colored skeletons buried next to the Red Queen represent Mars' two moons, Deimos and Phobos. If that hypothesis is accurate, then Palenque's temples tell the story of Mars—either from the past or the future. For instance, the "Palace" at Palenque seems to replicate a three-step, gravity-flow processing plant or chemical treatment facility, perhaps for water. While much larger, its layout and excavation terrain are similar to those of the Mars rover, "Curiosity."

Black Hole

According to the Mayan map, there may be a "Holbox" (Mayan-Yucatec for "Black Hole") near earth, as "Holbox Island" is near Chichen Itza on the Yucatan coast. Specifically, the waters of the island's Yalahau Lagoon represent the black hole. This may represent an "Isolated Horizon," since islands are isolated landmasses. See *"Satellite Photographs 'Black Hole' on Earth,"* at LiveScience.com for images.

The Pleiades

The temple layout at Tikal follows the same pattern as the Pleiades because they are what it represents. Thus, as the bloody events portrayed on the earth maps (Chichen Itza and Uxmal) indicated where those events happened (or would happen), so do the images and hieroglyphics at Tikal tell its story.

The Ramayana

The Mayan city of Copan, in Honduras, contains images that are very similar to that of Hanuman, the Hindu deity from the epic "Ramayana" (pronounced "Ra-*Mayan*'-a"). Hanuman was part human and part monkey. He was physically strong, wielded a deadly mace in his left hand, and was subservient to Rama. Thus, he was often depicted on bended knee.

In the image below (on the left) we see Hanuman. On the right we see an unnamed figure of similar description at Copan. He is both man and monkey. As the Ramayana describes, Hanuman's tail is damaged. In the epic it was set on fire by Ravana's men. In the image below, Hanuman is on bended knee, reflecting his subservience to Rama. With his left hand, this monkey-man holds a mace (a "gada" in Hindu mythology), and—as is common in Hindu imagery—his right hand is over his heart. This represents his devotion to Rama. The "T" on the mace represents death in ancient Mesoamerican culture. This means the object is a weapon rather than a tool. The larger image below shows Hanuman with his hands over his heart, as he shows his devotion to

Rama. If the similarities between the two images are not coincidental, then Hanuman may have been real in some sense—perhaps a reference to the hypothetical time travelers and the Hindu epics, which may reflect extraterrestrial phenomena.

LEFT: HANUMAN WITH MACE. PHOTO BY JEREMY RICHARDS | DREAMSTIME. RIGHT: MAYAN IMAGE WITH MACE AT COPAN. PHOTO BY ANTON IVANOV.

MAYAN IMAGE AT COPAN. PHOTO BY VALERY SHANIN.

Chapter 4

The Afterlife, Calendar, and Hall of Records

Mayan Afterlife

The Mayan afterlife is presented in the ancient city of Mayapan. According to experts it symbolically represents "New Chichen Itza," which—in the context of the map theory—means "New Earth." While the structures at Mayapan are all Pre-Columbian, the Chilam Balam of Chumayel indicates that they share symbolism with Christianity. For instance, one portion of it reads "…our Lord Dios…" (i.e. Christ) was worshipped "…according to the word and the wisdom of Mayapan." Similar imagery is found in the Codex Vaticanus, wherein the Aztec/Maya death god (who holds a spear) opposes the Aztec/Maya god-man who holds (what appears to be) a shepherd's crook over a mass grave/tomb. Moreover, archaeologists Marilyn Masson and Carlos Lope have identified Mayapan as the "realm" of Kukulkan—the god-man.

Mayapan is a version of St. John's Apocalypse written in stone and colorful paint. One enters the city through a wall with twelve gates. Magnificent beings stand along the wall at each of its four cardinal directions (not visible to tourists). From there, one passes the Temple of the Fisherman God, where a mural shows him in a fishing boat next to his enemy—a bound dragon. According to Mayan mythology (See *Kukulkan's Realm* by Masson and Lope), the Fisherman God resurrects the dead and gives them new bodies made from their old bones. Thus, dozens of real human corpses of men, women and children were found in the tomb section of the Temple of the Fisherman God. They may represent St. Paul's "elect."

At the base of Mayapan's Pyramid of Kukulkan, a series of murals shows a sacrificed man who shines like the sun. There are several other apocalyptic symbols. These include a woman at a grinding stone ("rapture" symbolism), incense holders, a pregnant woman, a "bottomless pit" where real human corpses were found among ashes, and more. See temple images.

TEMPLE OF THE FISHERMAN GOD, MAYAPAN.

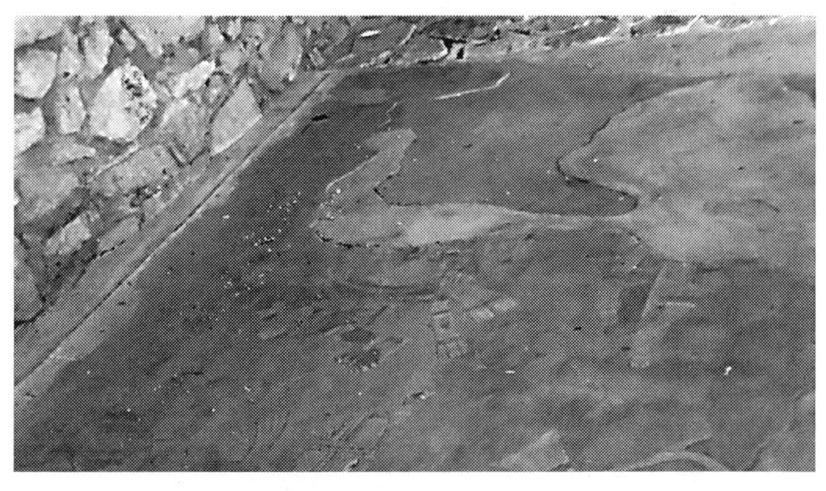

THE FISHERMAN GOD MURAL, TEMPLE OF THE FISHERMAN GOD, MAYAPAN IMAGE BY AUTHOR.

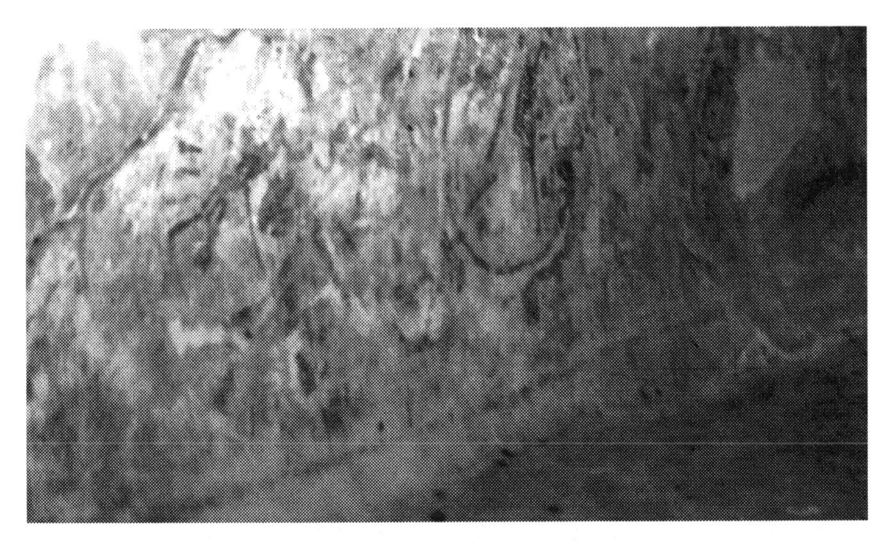

SACRIFICED MAN SUPERIMPOSED WITH THE SUN, PYRAMID OF KUKULKAN, MAYAPAN, IMAGE BY AUTHOR.

WOMAN AT A GRINDING STONE, PYRAMID OF KUKULKAN, MAYAPAN, IMAGE BY AUTHOR.

CENTER-LEFT BACKGROUND: TEMPLE OF THE BURNING PIT (AUTHOR'S TERMINOLOGY), MAYAPAN.

TOP VIEW: TEMPLE OF THE BURNING PIT.

Thus, apparently the Pre-Columbian Mayas (i.e. the architect) knew about Christian eschatology. Given that Mayapan has 4,000 detailed structures, the ancient Mayas may have known more about Christianity than did the Catholic priests who came to instruct them about it—often forcibly. Thus, de Landa's religious hegemony and destruction of Mayan books and statues ("idols") may have been counterproductive. Perhaps a sense of superiority prevented the Spanish priests from becoming both teachers and students of the Mayas, exchanging information on equal terms. Similarly, perhaps the Mayan Catholics continued practicing their former religion because they saw their two belief systems as complementary.

The Mayan Calendar and the Map

While many contemporary Westerners may have dismissed the Mayan calendar as an unreliable predictor (since the world didn't end in December 2012), it is proposed that its true purpose has been misunderstood. The Mayan calendar is not an apocalyptic alarm clock that rang too soon. Rather, in tandem with the Yucatan map, it hypothetically indicates where each planet (that corresponds with a Mayan city) was located on August 11th, 3114 B.C.—the *first* day on the Mayan calendar. Perhaps, then, the *last* day on the Mayan calendar offered the same planetary juxtapositions as the first day (a loop of sorts)—to orient us regarding the whereabouts of extraterrestrial intelligent life.

MAYA STELE SHOWING CREATION DATE OF 13 BAKTUNS, 0 KATUNS, 0 TUNS, 0 UINALS, 0 KINS, 4 AHAU, AND 8 CUMKU - AUGUST 11, 3114 BC.

Chapter 5

Freewill, Fate and Intervention

The map theory raises questions about man's place in the universe, freewill, and destiny. For instance, the architect either simply knew Abraham Lincoln would be elected, assassinated, and memorialized in Washington, D.C. a thousand years ahead of time; or he caused all that—including the city's design.

Clear thinking should normally dismiss foreknowledge and causation here. However, we should hesitate to be dismissive. According to author Jim Marrs, Sir Fred Hoyle (the Big Bang theorist) believed that scientifically advanced time travelers influence man like "pawns in a great game." That Hoyle would say such a thing rings true. He didn't trust scientists or governments. To Hoyle, scientists were "liars" who withheld potentially unpopular discoveries from the public for the sake of their careers; and politicians were no better. Thus, assuming Hoyle really did find evidence of time travel, then the architect has been *influencing* mankind. This could have been conscious or unconscious influence on our part, and it isn't clear which is worse. One puts world leaders in a secret cabal led by an odd advisor, while the other suggests mind control.

Conscious Influence

The cabal hypothesis implies a "blueprint theory." Meaning, if it is accurate, then the architect has been instructing world leaders—through the ages—regarding infrastructure, urban design, weaponry, and more. Accordingly, as Chichen Itza's Great Ball Court would represent the architect's blueprint for Washington, D.C., this would mean he "told"

American leaders where to place the National Mall, White House, the Smithsonian, the Vietnam Memorial, etc.

Unconscious Influence

The architect's alternative strategy of unconscious influence would mean world leaders—and perhaps average citizens as well—have been subjected to mind control. This would warrant reexamination of the "Montauk Project" allegations (a conspiracy theory about time travel and mind control) "Camp Hero," and the testimonies of Preston Nichols and Peter Moon.

Maya Religion

The map theory proposes that each Mayan city's temple: layouts, juxtapositions, rituals, bas-reliefs, statues, codices, myths, deities and terrain, were designed to tell the story of a corresponding planet or planetary cluster (i.e. a constellation). Thus, as Witschey and Brown (2011) estimate that there are 4,400 Maya sites, the corresponding number of planets with intelligent life falls within the Drake estimate.

According to the map theory, the primary duty of the ancient Mayan priests was to physically maintain the artifacts mentioned above, and to pass their meanings along to future generations of priests. For instance, while the architect designed the Tomb of the High Priest to tell the stories of Thunderbird dancing and "Kinkaid's Tomb" at the Grand Canyon, the priest's job was to explain the symbolism. However, any influence of the architect is still difficult to identify.

Chapter 6

Augustus and Alice Le Plongeon

Freemasonry

Augustus Le Plongeon was a Freemason and an early archaeologist of Chichen Itza. In search of the origins of Freemasonry, he and Alice believed they were the reincarnated ancient masonic couple—Prince Coh and Queen Moo. As supporting evidence for his past life (akin to genetic memory) hypothesis, Augustus identified the location of what he claimed to have been a statue of himself—25 feet below Chichen Itza's Great Ball Court. He could not have observed the statue prior to digging it up. Yet, he knew where to dig.

Lost Support

At first all the adepts marveled at the Le Plongeon's archaeological findings. Alice was taking photographs, and they were writing books—as planned—but, then their work became more esoteric, and the scientific community withdrew. It was an unavoidable separation, as Augustus turned from practicing good science to fulfilling his masonic mission. Still, however, he seemed less interested in arcane rituals and secret handshakes, and more interested in the unknown.

LEFT: DR. AUGUSTUS LE PLONGEON IN MASONIC REGALIA.
RIGHT: ALICE DIXON LE PLONGEON (WITH COMPASS).

In other words, Augutus wasn't looking for the origins of Freemasonry in the usual sense of Albert Pike's symbols, or concepts in line with Adam Weishaupt or Manly Hall on an ancient rock. He wanted more. Peter Tompkins described Augustus as being privy to ancient Mayan secrets, and as one who:

> "...could speak of cycles of existence in more advanced planets and worlds than the present..."

The Great Architect of the Universe

Tompkins believed Le Plongeon had accessed extraterrestrial information, presumably in the Yucatan, which means the masonic "Great Architect of the Universe" may be the human who designed the Yucatan Hall of Records—the hypothesized time traveler.

THE GREAT ARCHITECT OF THE UNIVERSE
(WITH COMPASS).

If the Great Architect is the architect of Chichen Itza, then, ironically, Augustus and Alice may have been stumbling *through* their objective (the secret of the Great Architect of the Universe), while *searching* for it.

There would have been two reasons for the Le Plongeon's oversight. First, they were born too soon. Several of the time-anomalous structures (e.g. the Washington, D.C. memorials) in the hall had not yet been built. Second, the Le Plongeons excavated Chichen Itza before the Wright Brothers flew at Kitty Hawk. So, there would have been no bird's eye view that would have shown the Western Hemisphere's familiar contours. Therefore, it didn't occur to them that Chichen Itza (and Uxmal) were maps of earth, or that—by extension—the other Mayan cities represented other planets.

Chapter 7

Finding the Hall of Records

Pedestrian Map Observability

Architecturally, the Mayan maps are like the walkable map at Federal Plaza in Washington, D.C. However, they differ in dimension, size, and scope. The Washington map is one square block and represents a city, while the Chichen Itza map is roughly one square mile and represents the Western Hemisphere in three dimensions. One clue that Chichen Itza is a map of earth (the West), is that its main entrance is located where the Yucatan Peninsula would fall on our modern maps. Entering visitors walk east, just south of the outcropping that represents Florida. The secondary entrance, near the Mayaland Hotel, is modern.

The Implausibility of a Hall of Records

The Hall of Records is considered to be a highly implausible concept, and few serious researchers have considered it. An ancient Egyptian scroll indicated a Hall of Records in Egypt, and the psychic Edgar Casey said there are three halls—in Egypt, Bimini and the Yucatan. However, ancient legends and psychics are not considered to be valid information sources. Perhaps part of the hall's implausibility stems from the common belief that the "big questions" (i.e. extraterrestrial life, etc.) have no apparent answers.

The Implausibility of Time Travel

One of the criticisms of backward time travel is that it lacks evidence. If it were possible—the argument goes—then why hasn't someone from our future visited us… and left evidence? This argument assumes the time traveler's direct audience would have to be us—modern man. However, metaphorically speaking, what if the football (time travel evidence) had been thrown over our heads to someone behind us—to our ancient ancestors? If we rule that possibility out, then we risk misinterpreting ancient evidence. Consider, for instance, the Davenport, Iowa Tablets.

The Davenport Tablets

If Fred Hoyle was right, and there are time travelers among us, then we may have been dismissing or distorting the evidence of their existence. For instance, aside from the modern aircraft images at the Temple of Seti I in Abydos, there is evidence on a set of Egyptian tablets that were excavated in Davenport between 1876 and 1877. In ancient symbolism, the tablets show an ideologically-driven man attacking by air. He does not use his left hand, and he is about to die. The tablets also show twin towers collapsed on September 11th, the mass cremation of "captives," four flying objects, and other indicators of 9/11. Harvard University's Professor Emeritus, Dr. Barry Fell, said the tablets were carved by a Libyan astrologer-priest of Osiris named "Wnty" ("Star-watcher") who sailed up the Mississippi River in 800 B.C. Hence the Egyptian symbolism on the tablets. The amateur archaeologist who found the tablets (in a Native American Indian mound) was first regarded as the hero who had found "America's Rosetta Stone." However, subsequently, the public saw him as a chump who had been tricked, after which the story takes an odd twist.

James Wills Bollinger, a respected Iowa judge (and member of a club called "The Immortals") announced that he had carved the tablets when he and his friends were drunk at a brothel. The creation process, he explained, entailed copying random symbols from the Encyclopedia Britannica onto shale slabs, and then burying the tablets in an Indian mound. The goal had been to prank Rev. Jacob Gass, a local amateur archaeologist.

LEFT (DAVENPORT STONE #1 SIDE A): TWIN TOWERS FALLEN ON THE FIRST DAY OF AKHET (SEPT. 11TH BETWEEN 1958 AND 2098). RIGHT: (STONE #1 SIDE B): MASS CREMATION SCENE.

The case took another twist, however, because there are temporal oddities with Bollinger's "confession." First, by all accounts (including that of the Iowa State Archaeologist) Bollinger would have been only nine years old at the time. Second, according to experts (e.g. Peter Tompkins), no one on earth knew how to translate all three of the tablet's languages in 1877. Third, Tompkins said the encyclopedia volume that Bollinger allegedly consulted contained no such information. Thus, if Bollinger really did carve the tablets, then he used an information source unknown to the leading anthropologists and language experts of his day. Furthermore, it would mean Bollinger accurately presented the 9/11 story by chance, and buried it (on tablets) at the only other place in America (besides Manhattan) where a large island rests in the middle of a large river at 41 degrees north latitude. However, on the other hand, it isn't clear why a prominent judge would lie about such a thing.

While it would not have been impossible for a nine year old to have gone out drinking and carousing with his friends at a Mississippi River "cat house" in 1877—after imagining real symbols from ancient Punic—and then using them to explain how to construct a calendar out of sunlight and a mirror on the spring equinox, it is unlikely. Either Wnty or Bollinger was an excellent guesser, or one of them is in the center of a temporal anomaly involving 9/11 and the Davenport

Tablets. There is another temporal connection, however. When Michael Drosnin—author of The Bible Code series— "asked" the software for the missing code key, it described the Davenport Tablets (i.e. in multiple languages, two tablets, two obelisks, next to a corpse, in a temple, in an archaeological mound, on Cook Farm, in a Valley of Limestone, etc.). According to Drosnin, the matrix also said the code key would be delivered by a "Time Traveler" with a "Time Machine." My own research of Drosnin's Hebrew matrix places the word, "Pacal" adjacent to "Time Machine." Thus, the Torah Code seems to have attributed the Iowa tablets to a time traveler. More details and images are available in my previous book: *Time Traveler in the Age of Aquarius,* about a temporal anomaly involving 9/11 and the Davenport Tablets.

The Orient Point Stone

Supporting evidence for the Davenport Tablets' authenticity was provided a few decades later. On New Years' Day 1920, a man named Daniel Young delivered a small stone to the Native American Museum in New York City. He said he had found it many years prior, while digging for oysters at Orient Point (the tip end of Long Island on the opposite fork from Montauk). The stone's Egyptian hieroglyphics mentioned sailors from Egypt, which Dr. Barry Fell (the Harvard professor who had dated the Davenport Tablets to 800 B.C.) said had been carved in roughly 900 B.C. Subsequently, the Smithsonian Institution acquired the Indian Museum. However, when they took inventory in 1976 the Orient Point Stone was missing. It still has not been found.

Additional supporting evidence for the Orient Point Stone (and therefore for the Davenport Tablets' authenticity) is what appears to be ancient Egyptians living on Long Island. Consider, for instance, Rufus Wilson's historical account of Long Island, which includes a 1902 photo called, "The Pyramids of Montauk." It shows three small pyramids roughly 20 ft. tall. Consider also that the Montaukett Indian family name of chiefs is "Pharaoh." Moreover, the name "Montaukett" is phonetically similar to "Mintaka," one of the stars in Orion's belt that was replicated among the pyramids of Giza.

Stories of an unusual Indian language being spoken at the tip end of Long Island captured the attention of Thomas Jefferson, who subsequently went there to investigate. Then, supposedly, however, a gust of wind blew all his notes into the Potomac River, thereby destroying all evidence of this now extinct language.

LEFT: STEPHEN TAUKUS PHARAOH OF THE MONTAUKETT INDIAN NATION, RIGHT: *PYRAMIDS OF MONTAUK* (ORIGINAL TITLE), *LONG ISLAND HISTORY*, BY RUFUS WILSON (1902), PYRAMIDS 2 AND 3 ARE LEFT AND BEHIND PYRAMID 1. NOTE: FOR MORE INFORMATION ABOUT THE MONTAUKETT INDIAN NATION SEE BOOKS BY EVAN PRITCHARD.

The existence of the Orient Point Stone is important because, since Bollinger did not carve it (or he would have said so), then it is evidence of ancient Egyptians in America, which means they might have carved the Davenport Tablets—not Bollinger. However, like a Sherlock Holmes mystery, the story has one more twist—"Project Pegasus."

Project Pegasus

According to Washington State attorney and 2016 U.S. presidential candidate Andrew Basiago, Project Pegasus involved governmental time travel experiments. Basiago claims to have participated in them when he was between the ages of 7 and 12, which was Bollinger's age when he would have carved the Davenport Stones—which means maybe Bollinger was telling the truth after all. "They trained children..." Basiago explained to the Huffington Post in 2012, "...so they could test the mental and physical effects of time travel on kids…" He added that children are physiologically more capable of adapting to the rigors of time travel than adults, and that a recent American president had been one of his fellow "chrononauts." He also claimed to have remotely viewed past events (from a few thousand years ago) with the aid of a modern device called a "chronograph." In the image below, Basiago claims, the U.S. Government had dressed him as a Union Soldier Bugle Boy to surreptitiously attend one of Abraham Lincoln's speeches.

However believable or unbelievable Basiago's story and his alleged photo near Abraham Lincoln are, the Iowa tablets are less questionable. They are on display at the Putnam Museum in Davenport; the major 9/11 facts are undisputed; the events of 9/11 are observable; and the tablets preceded the attack. But, let us continue discussing the reasons for the delay in finding the Yucatan Hall of Records.

LEFT: LOWER CENTER - ALLEGED IMAGE OF ANDREW BASIAGO, WITH ABRAHAM LINCOLN INSIDE THE SQUARE. RIGHT: IMAGE OF ABRAHAM LINCOLN.

Missing Data

Vandals, treasure hunters, museum suppliers, etc. destroyed or removed many of the structures from Chichen Itza, which has impeded the map's identification. For instance, had the "Ireland" temple been destroyed, it would have been more difficult to identify Europe. The temples that represented Sub-Saharan Africa were destroyed to make space for the Mayaland hotel.

Similarly, a museum in Merida displays a tablet from Chichen Itza's El Caracol, and the information card next to the tablet indicates a specific time, person, and event. El Caracol represents Fajada Butte in Chaco Canyon, NM, a sacred place to Native Americans. However, since the tablet was removed, it is difficult to interpret its messages because we don't know the precise spot where it was found. Such destruction is the equivalent of removing Antarctica from the original Piri Reis map.

Fortuitous Timing

Previous generations of researchers could not have identified the Hall of Records because many of the temples that comprise the map had no real-world correspondents, as explained previously regarding the Le Plongeons. For instance, no one could have figured out that the Temple of the Bearded Man represented the Lincoln Memorial prior to the Lincoln Memorial's construction. The same holds true for the White House, the Vietnam Wall, and the English Channel Tunnel replicas. Similarly, while the map shows a Stonehenge on a sunken island west of Sicily, it was only in 2015 that archaeologists found the Stonehenge monoliths on it.

Temples vs. Tablets

Contrary to popular belief, the Hall of Records is not many tablets in a temple, but many temples that contain relatively few tablets. Most of the information is inscribed on walls rather than on tablets.

Circular Logic for Temporal Anomalies

Anecdotal evidence suggests that time travel evidence is often processed with circular logic. For instance, one might conclude that the English Channel Tunnel replica cannot represent the actual tunnel, because that would require time travel (assuming clairvoyance has been ruled out). But, since time travel does not exist, the tunnel replica cannot be evidence of it. It is like concluding that since birds don't exist, the animals that fly must not be birds.

Consider, for example, the Temple of Seti I in Egypt, which seems to show a helicopter, boats, and a jet, alongside an insect and a fish. One can interpret the boats etc. as evidence of future knowledge. Or, one can interpret them as evidence of "water damage," as some archaeologists have hypothesized.

An American Alert

The water damage hypothesis is understandable, as some Egyptian messages have been hidden. For instance, the sacred "Benben Stone" shows a gorilla in a space capsule replica, but "Figure & Ground" conceals his face. As he was accompanied by the "God of Time," the message is about a modern day "space monkey" in ancient Egypt. Assuming he represented an American effort, that would explain the Davenport message from an ancient Egyptian "priest" about twin towers collapsing on September 11th, a fiery suicide attack by air, a mass cremation, a large tree in a holy place (i.e. the "Miracle of the Sycamore" at Ground Zero) etc. The gorilla is discussed in more detail later on.

Chapter 8

Validity

Borrowing from standard approaches to measuring validity, this discussion suggests several requirements for qualifying a document as a map. While the following definition is flawed, I am basically defining a map as "any manmade object that intentionally functions as a map."

Maps are Intentional

A map should appear to be an attempt to model geographic reality. Thus, if Chichen Itza is a map, then it should appear to be an attempt at one. It should not offer the interpretive flexibility of a Rorschach test. Evidence that Chichen Itza resembles the Western Hemisphere can be found in satellite images of the ruins, particularly at the Southern Europe portion of the map.

Maps Have Discriminant Validity

Maps must discriminate between locales. For instance, a map that describes New York just as well as it describes Mumbai is too ambiguous, and does not qualify as a map. In the instance at hand, the Maya's Western Hemisphere map does not (and should not) also work for the Eastern Hemisphere. Similarly, the earth map (Chichen Itza and Uxmal) should not also describe other planets.

Maps Have Criterion Validity

A bona fide map should offer criterion validity. Culture, style, and artistic differences aside, it should resemble other maps of the same places. For instance, France on Map A should clearly correspond with

France on Map B, and so on. The only discrepancies should be due to style. For instance, the Maya style lies somewhere between abstract and impressionist. Moreover, objects on the Mayan map do not follow the Western style of precise coordinates. Neither are they scaled. The reason is that the Mayan map uses alignments to present messages. Therefore, natural, coincidental, alignments are apparently deliberately misaligned. For instance, the Azores replica (the Pyramid of Kukulkan) is farther north than it should be to align it with Portugal, and to place it over a cenote as explained earlier.

Similarly, the Iceland replica is farther south than it should be. This represents Iceland's cultural proximity to Western Europe, and its cultural distance from Inuit culture in the Arctic. Moreover, the Portuguese Algarve replica on the Chichen Itza map resembles the actual sea caves at the Algarve. They are oversized, however, to indicate their importance as a landmark between the Atlantic Ocean and the Mediterranean Sea.

Maps are Internally Consistent

Maps should be internally consistent. The French mainland, for instance, must be north and west of the Mediterranean Sea. Furthermore, it must be south of England and east of the Atlantic Ocean, etc. Similarly, in turn, the Atlantic Ocean must contain few landmasses and be longer (north-south) than it is wide (east-west). In addition, Africa must be on the east coast of the Atlantic Ocean, and the America's must be on its west coast.

Maps Offer Geographic Details

A map must offer enough detail to distinguish one location from another. For instance, an empty circle does not qualify as a map of a planet—unless it is shown in the context of other planets. Similarly, a sufficient number of landmarks qualify a document or structure as a map.

Chapter 9

The Mayan Map's Legend

According to anthropologists, there are no fixed linguistic rules in Mayan hieroglyphics. This applies to the map structures as well. Thus there are exceptions to the following rules.

Ball Courts: The Maya word for ball court, "hom" means "crevice in the ground." Technically, it refers to a crevice in the back of a large turtle. Thus, ball courts represent fault lines and their earthquake zones, whose tectonic plates creep along slowly, like turtles, but strike quickly and without warning. Ball courts with high parallel walls represent amplified sound (e.g. Chichen Itza's Great Ball Court). Ball Courts set at right angles indicate a planet with multiple light sources (e.g. suns). The logic is that the courts at Chichen Itza run north-south so that no team is at a sunlight (i.e. blinding) disadvantage. Thus, the absence of a ball court may mean the corresponding planet has no sun.

The angle of a ball court's inclines (aprons), their length and width, and the width of the playing field (trench) all indicate characteristics of the fault zone. Presumably, relatively horizontal inclines indicate earthquake epicenters closer to the surface of the ground, and steeper aprons indicate epicenters closer to the earth's core. Wider ball court spaces (between aprons) may represent wider the fault lines.

MAYA BALL COURT (AT COBA).
PHOTO BY JAVARMAN | DREAMSTIME.

Ball Court Rings: are cross-sectional planetary earthquake maps. Rings (see below) with intertwining snake motifs or parallel lines (not shown below) indicate shock waves. Tongues of attacking snakes meet at the perimeter of the ring to indicate the focal point of the earthquake. The diameter of a ring's empty center may indicate the size of the corresponding planet's inner core. The width of the solid portion of the ring may indicate the thickness of the planet's crust.

Contemporary seismologists employ similar earthquake models. Like the Maya's, theirs include four concentric rings to indicate earthquake shockwaves. Their "Focus of the Earthquake" in the image below is where the Maya model shows twin snakes striking their victim—the surface of the earth. Thus, as mentioned previously, the opposing Maya ball court teams represented opposing tectonic plates—not opposing tribes. As will be shown later, the Chichen Itza and Uxmal maps pinpoint earthquakes at Crete, Morocco, the Adriatic Sea, Washington, D.C., and Qinghai, China.

The ritualized execution of the winning Maya ball court captain represented earthquake casualties. The following logic of executing the winner is offered: The first team to score (by knocking the ball through the center of the ring), won the game. However, in actuality, if someone is close enough to a fault line (during an earthquake) to drop an object into it, then he will probably be killed. Accordingly, the captain was executed to symbolize the death of his team, and sometimes multiple players were executed. While it may be disturbing to think of it this way, this is simply more pedagogic theater. The lesson—presumably the work of the architect—is that earthquakes are deadly. Moreover, they were to occur in specific locations along the Hellenic Trench, the Adriatic Plate, etc.

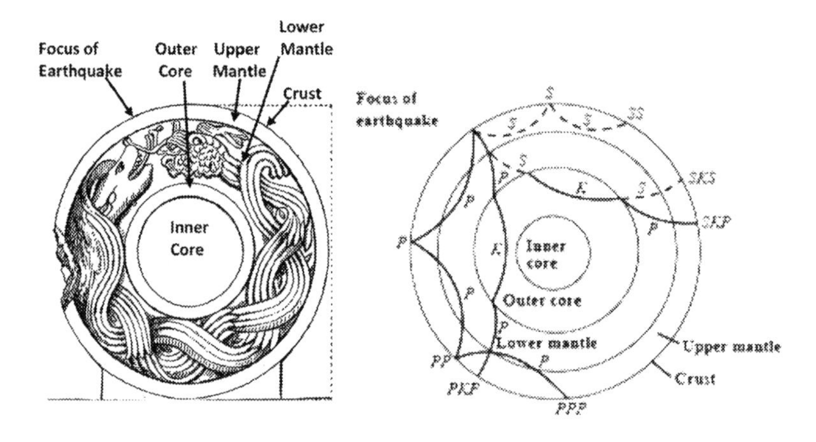

LEFT: SERPENTS AS PRIMARY AND SECONDARY SHOCKWAVES, GREAT BALL COURT, CHICHEN ITZA BY CATHERWOOD (1843). RIGHT: MODERN SEISMIC MAP OF EARTHQUAKE WAVES-UNITED STATES GEOLOGICAL SURVEY.

Cenotes: These are naturally sunken lakes in sinkholes. They contain lighter-weight fresh water on top, and heavier sea water below. Wet cenotes represent large continental bodies of water (e.g. the Southern Cenote of Xtoloc as "Lake Chad"), and dry cenotes represent dry valleys (e.g. Akab Dzib, symbolizing cliff dwellings).

Chaac: the rain god indicates where the map requires water (i.e. to create rivers, seas, etc.). For example, Chaac appears on the replica of Libya's Mediterranean Sea coast.

Chac Mool: seems to be a zeitgeist—a Hegelian "spirit of the age" who acts as the map's storyteller-guide. He connects disparate parts of the map (for observers) by having us follow his line of sight and study his facial expressions. For instance, a Chac Mool looking up (on the map) means there is something important to be seen in the sky at the corresponding location on the relevant planet. When he is swimming or looking impoverished, he is presumably indicating water and drunkenness or poverty. Similarly, when he is sensually aroused on some part of the map, it means something sensual is connected with the corresponding locale in actuality.

Accordingly, the Chac Mool in the replica of the volcanic Mt. Pico represents long ago, when ancient Azoreans worshipped nature. Similarly, the Chac Mool in the Notre Dame replica shows him looking west—across the Atlantic Ocean—indicating French-Catholic colonialism (e.g. Quebec, Louisiana, Haiti, etc.). There was also once a Chac Mool near the Grand Canyon replica (the Tomb of the High Priest).

Consistent with the Chac Mool-as-guide hypothesis, Peter Tompkins suggested that Chac Mool's lap-dishes once held compasses—magnetic pointers floating in water or liquid mercury. It is difficult to determine if Chac Mool instructs the map visitors, or if he represents unseen forces that influence mankind.

A CHAC MOOL STATUE FROM CHICHEN ITZA (1800'S).

A CHAC MOOL STATUE AT THE MUSEO NACIONAL DE ANTROPOLOGIA.

Hunapu and Xbalanque: are the archetypal heroic team and enemy of Vucub Caquix, the archetypal tyrant. On the Chichen Itza map they represent George Washington and John Adams. This is discussed in detail later on.

Infant Burial Grounds: represent burial grounds for victims of ritualized infanticide. Portions of graveyards containing real infant corpses in clay jars were found in the African and South American replicas at Chichen Itza. Similarly, there is contemporary evidence of infanticide in Sub-Saharan Africa, and among some Amazonian tribes in Brazil.

Kukulkan: is a feathered (meaning—from/of the heavens) serpent (meaning—from/of earth) deity. In the map theory, not every snake is a Kukulkan, and not every Kukulkan is a deity. He seems to usually represent a deity, and perhaps may sometimes represent a human ruler.

- *Right Angle Kukulkan*: A Kukulkan with a vertical body and a horizontal head represents "A.D." (Anno Domini), "Christianity," or both. The rationale is that the right-angle Kukulkans appear at the Easter resurrection display at the "Upper Temple" of the Pyramid of Kukulkan. The right-angle Kukulkans also appear at the Notre Dame replica on top of the Temple of the Warriors.
- *Slanted Kukulkan*: A Kukulkan with an angled body (roughly 45 degrees), and a horizontal head represents "B.C." Examples appear at the middle portion of the Pyramid of Kukulkan and at the Tomb of the High Priest.
- *Horizontal Kukulkan*: A fully horizontal Kukulkan represents influence (e.g. government). Examples include the Platform of Venus and the Platform of Eagles and jaguars.

Pakal: Men named "Pakal" (also "Pacal") appear in more than one Mayan temple city. There is one carved on a sarcophagus in Palenque, and another carved on a stele that once stood in the Southwest U.S. replica at Chichen Itza. It shows a hornbill duck among a set of hieroglyphic inscriptions. Since there are no hornbills in the Southwest U.S., presumably the bird refers to a flying traveler.

Pakal means "shield" in English. It is the term American astronauts and military personnel use for their insignia. Ancient astronaut theorists interpret Pakal's sarcophagus as a man in a spacecraft, while archaeologists interpret it as Pakal traveling to the afterlife. However, they are both right. The sarcophagus shows Pakal traveling between two worlds, but they are earth and Mars. In other words, "Pakal" is a not a name, but a job title meaning "Space Traveler with Insignia," "Flying Traveler with Insignia," "Captain," etc. This will be made clearer later on when the sarcophagus is presented. It is a three-part puzzle that shows "Lord Pakal" ("Lord Shield") as an American astronaut to Mars.

SHIELD/INSIGNIA OF THE U.S. FEDERAL AVIATION ADMINISTRATION.

Pillars: represent pillars and individual people. Many of the pillars in the Mediterranean Sea portion on the map are engraved with human images. They may represent society in general, or specific people.

Platforms: represent isolated structures with barriers. At certain times of the year and day they produce an eagle and a serpent in stone and shadow media. Since they are on a stone, they symbolize an "island" in Mesoamerican culture. Examples include the Iceland (Platform of Venus) and White House (Platform of Eagles and Jaguars) replicas.

Plazas: represent large intercontinental waters—oceans and seas. Chichen Itza's Grand Plaza represents the Atlantic Ocean.

Pyramids: represent mountains or tombs. For example, the pyramid Tomb of the High Priest represents the tomb Kinkaid found in the Grand Canyon. The pyramid beneath the Pyramid of Kukulkan represents Mt. Pico in the Azores.

Replicas: Unique map structures ("replicas") resemble their correspondents in appearance and/or function. For instance, El Caracol resembles Fajada Butte in Chaco Canyon. In many instances the replica's construction preceded that of the "original." For instance, the replica of the Lincoln Memorial was built a thousand years before the actual Lincoln Memorial was built. However, Edgar Casey's "New Atlantis" theory might suggest that Washington, D.C. is the replica (not Chichen Itza's Great Ball Court)—of the ancient capital city of Atlantis—and that the Mayan map simply records that connection. Such a hypothesis would be based on Casey's theory that ancient Atlanteans reincarnated as Americans.

Sacbes: These ancient "White Paths" represent planetary magnetic fields. Sacbes between Mayan cities represent Interplanetary Magnetic Fields.

Tenoned Snakes: Tenoned snakes with wings represent people who are inserted tightly into flying objects. Therefore, they represent aircraft pilots.

Vucub Caquix: (aka "Seven Macaw" or "Seven Parrot") represents King George 3rd of England. Any allegorical reference to him places the story in either England or America. For instance, the story of Zipacna and the "Four Hundred Drunken Men" at the Great Ball Court refers to England (since Zipacna was the "son" of Vucub Caquix) and Washington, D.C. (the U.S.).

Walls: Low, *narrow* stone walls may represent major rivers. The Yellow Nile replica is presented this way. Low, *massive* stone walls represent continental coastlines. The Atlantic coastline replicas are depicted this way. A low, long wall was also used to indicate the Arctic Circle.

Xtoloc: ("Esh'-toh-loke" meaning "Iguana") temples represent societies near large bodies of water (e.g. cenotes) in areas with extreme heat or cold.

The Eastern Hemisphere

Chapter 10

The Eastern Hemisphere

Partial Map of Earth's Eastern Hemisphere at Uxmal (900 AD)

NORTH GROUP BY HJPD. DOVE HOUSE AND SKULL & BONES BY V. MURRAY. GOVERNOR'S PALACE BY MESOAMERICAN. UXMAL PHALLI BY CHICO SANCHEZ. SITTING FIGURE (UXMAL) BY MCDUFF EVERTON. 1 = RUSSIA, 2 = BUDDHIST TEMPLE IN BEIJING, 3 = FORTUNETELLING IN BEIJING, 4 = CENTRAL TIBET, 5 = LHASA, TIBET, 6 = THE HIMALAYAS, INDIA, AND NEPAL, 7 = THAILAND, 8 = MT. EVEREST, 9 = SRI LANKA, 10 = CAMBODIA/ANGKOR.

Before we discuss the map's Western Hemisphere (Chichen Itza) in detail, let us briefly discuss the east. The Eastern Hemisphere is presented in the ancient Mayan city of Uxmal. It is located about 120 miles southwest of Chichen Itza. An actual map of the Eastern Hemisphere is not provided here because Uxmal doesn't include landscapes—only landmarks. However, each landmark has been verified with Google Maps.

Comparing East and West

The difference between the Eastern and Western Hemispheres on the Mayan maps is that the East is characterized by "squares" (i.e. "The Nunnery" as Tiananmen Square) while the West is characterized by pillars (Greco-Roman influence).

India and Nepal

Pigeon House ("House of the Doves") is a thin wall of nine peaks which represents the world's nine tallest mountains in the Himalayas—a thin wall of high peaks. Archaeologist John Stephens' (ca. 1840) account of Pigeon House also accurately describes the Himalayas. He wrote, "… ascending the great staircase to the building on top… It commanded a view of every other building, and stood apart in lonely grandeur, seldom disturbed by human footsteps."

Pigeon House has three parts; the Southern Courtyard—still largely unexcavated—represents India. The house's nine peaks represent the Himalayas, and the Northern Courtyard (including the stairs) represents Nepal. In the image immediately below, India is left of the peaks and Nepal is to their right. Note: while I only counted eight peaks at Pigeon House, older sources counted nine. It isn't clear why the structure wasn't designed with eight peaks, which would make the Great Pyramid the ninth peak. However, rarely are structures at Uxmal comprised of an even number of sections. They are typically odd-numbered.

White doves perched on the Mayan structure presumably represent snow in the Himalayas. Evidence that the architect knew the structure would attract birds is Uxmal's "House of the Birds," where many birds nest among stone carvings of pigeons and other birds. Thus, apparently the architect used doves to make Pigeon House resemble the Himalayas. Moreover, Pigeon House peaks are see-through, which enables the observer to see white clouds—representing snow—in the background. Presumably, a millennium of weathering and damage has dulled Pigeon House's peaks.

HOUSE OF THE DOVES, UXMAL.
PHOTO BY RALF BROSKVAR | 123RF.

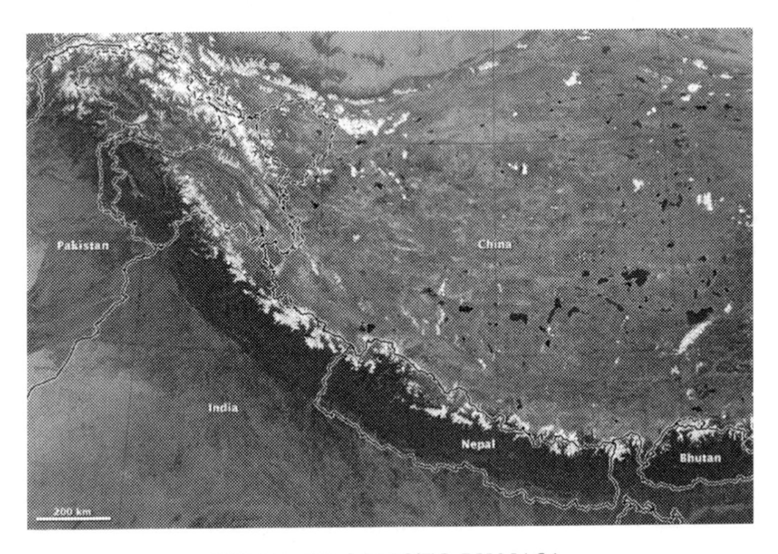

HIMALAYAS PHOTO BY NASA.

HOUSE OF THE PIGEONS, UXMAL, PHOTO BY AUTHOR.

THE HIMALAYAS BY DCHAUY.

Mt. Everest

Uxmal's "Great Pyramid" represents Mt. Everest (Sagarmatha or Chomolungma). Evidence of its status is the parrot on the North Face.

As parrots can represent kings in Mayan mythology (e.g. Vucub Caquix), so does the Great Pyramid's parrot identify it as the "king" (tallest) of mountains. That is why the north faces on both the map structure and the actual mountain look so similar. The grotesque faces on the pyramid may represent the perils of climbing Mt. Everest.

The Great Pyramid is adjacent to the east end of Pigeon House (the Himalayas replica), just as Mt. Everest is on the east end of the Himalayas. The Great Pyramid's north side has a rectangular box-like replica of an actual North Face ledge including a steep drop. The box appears to represent the start of the "Death Zone," above which the air is too thin to support human life.

Stone images of Asian men in knitted hats and parkas—at the summit of the Great Pyramid—represent Nepalese Sherpas. They are an ethnic group of skilled mountain climbers and guides. Aside from earning a living, they often climb Everest for spiritual reasons, rather than as a test of physical endurance. Their work is dangerous, and fatalities are not uncommon. Thus, the images of the men on the pyramid include only their heads, which may represent loss of life. Two of the men's images appear below.

Another image (not shown below) shows a sad-looking (non-Asian) man who is buried up to his neck. Falling parrots above him symbolize the mountain's deadly icefalls, the most popular of which is Khumbu Icefall. In one instance, without warning, a twelve-story block of ice fell over and killed a Sherpa.

THE NORTH FACE OF THE GREAT PYRAMID OF UXMAL. PHOTO BY JAVERMAN.

THE NORTH FACE OF MT. EVEREST, PHOTO BY YANG JUN.

IMAGES OF ASIAN MEN WEARING KNITTED HATS AND PARKA HOODS, TOP OF THE GREAT PYRAMID, UXMAL. PHOTO BY OLIVER DAVIS.

Chapter 11

Tibet

Lhasa: Potala Palace

The "Governor's Palace" at Uxmal is a replica of Potala Palace in Lhasa, Tibet. The Tibetan palace rests on 3,700 foot Marpo Ri ("Red Mountain"), which sits on an 11,450 ft. plateau. Thus, the Tibetan palace is on a hill, on a hill. Similarly, the Governor's Palace at Uxmal rests on a small hill on a larger hill.

Potala Palace is the Dalai Lama's traditional seat of government. While no structural features distinguish its Mayan counterpart as a seat of government, clearly "Governor's Palace" conveys the same meaning. Since local Mayas could not have known about Lhasa, or what transpired there on their own, the architect presumably told them long ago. This is consistent with the name "Uxmal," which means "the future."

Local Mayas refer to the massive stone in front of the Governor's Palace as a "Whipping Post." It may refer to the cruel whippings levied upon Tibetan peasants by Lhasa's wealthy lords. Thus, overall, the Mayan and Tibetan structures have similar names, functions, architecture, hilltop locations and map juxtapositions.

POTALA PALACE AND SHOL (LOWER), LHASA, TIBET. CHINAWDS | DREAMSTIME.

GOVERNOR'S PALACE, UXMAL. PHOTO BY ALEXANDRE FAGUNDES DE FAGUNDES.

The sculpture over the main entrance to the Governor's Palace presumably represents an image of Buddha at Potala Palace. The twin "lions" on the east side of the Governor's Palace represent the two snow leopards who guard the east gate of Potala Palace. The felines on the map are conjoined to indicate that they are twins.

The Tibetan snow leopard represents fearlessness, cheerfulness, the east, and earth. It is one of the four dignities of Shambala Buddhism. Twin leopards also serve as Tibet's symbol (e.g. on the flag). The Mayan map replicates the high peak on the Tibetan flag, and the peaks on the map and on the flag are aligned with the cats. However, when Le Plongeon discovered the Uxmal peak a century ago, he would have had no way of knowing it represented a mountain on the other side of the world.

TWIN JAGUARS AT THE GOVERNOR'S PALACE, UXMAL.

TWIN SNOW LEOPARDS ON TIBETAN FLAG.

Lhasa: The Parable of the Arrow

Tibetan archery has a long history in Lhasa, which might explain the arrow-shaped areas at the Governor's Palace in Uxmal. Images below include Alice Le Plongeon (and two workers) during excavation of the Governor's Palace. In the Le Plongeon image, the short stem of the arrow has not yet been excavated. Presumably the arrow at the replica of Potala Palace (the governor's Palace) is less a reference to archery, and more about the Buddhist "Parable of the Arrow." It means if you're shot with a poison arrow, your priority should be its removal—not investigating its designer, etc. In other words, sometimes the expected value of perfect information is negative. Ultimately, the parable relates to the "Fourteen Unanswerable Questions," the nature of the cosmos, life after death, etc.

ALICE LE PLONGEON AND MAYA WORKERS INSIDE
"ARROW HEAD" AT GOVERNOR'S PALACE.

GOVERNOR'S PALACE, ARROW FIGURE, PHOTO BY MOFLES.

Lhasa: House of the Turtles

The House of the Turtles is a rectangular structure immediately northwest of the Governor's Palace. As turtles live in water, so does this structure represent Zongjiao Lukang Park Lake. It is in the shape of a rectangle and is located a few hundred feet northwest of Potala Palace. Presumably, the architect did not use a cenote on this part of the map (to indicate a body of water) because—for geological reasons—cenotes are either rare or nonexistent on hilltops.

LEFT: HOUSE OF THE TURTLES, UXMAL,
RIGHT: ZONG-JIAO-LU-KANG PARK,
© LIZLEE | DREAMSTIME.COM.

Lhasa: Shol

At the base of Potala Palace's Marpo Ri hill lies the town of Shol (or "Zhol"), meaning "below." It is surrounded by walls on three sides, one of which is the palace side. Shol is the home of local people, palace administrators, and Shol Prison. The prison became infamous during the 1950's when photographs emerged of prisoners wearing large wooden collars that prevented them from touching their mouths. Thus, they relied on monks to feed them.

On the Mayan map, Shol is the small stone town behind and below the Governor's Palace. It is essentially down in a pit, but it accurately presents the spatial relationship between the actual Potala Palace and Shol. The Shol replica is surrounded by three walls, one of which is adjacent to the Governor's Palace—as we find at Potala Palace. The juxtapositions between the Uxmal and the Tibetan structures (i.e. the Governor's Palace and Potala Palace) and the towns "below" are nearly identical. Notice, for instance, the similarity between the image on the left, and the lower-left corner of the image on the right.

LEFT: A PORTION OF THE TEMPLE BELOW THE GOVERNOR'S PALACE, UXMAL. RIGHT: POTALA PALACE (TOP) AND SHOL (BELOW). IMAGE BY EVERETT HISTORICAL.

Central and Western Tibet

Uxmal's "Cemetery House," with its eerie skull and bones images, represents Central and Western Tibet. "Among the many customs of Tibet…" wrote Chicago Field Museum Curator of Anthropology, Berthold Laufer in 1923, "…none has attracted wider attention than the use of human skulls and other bones for practical purposes and in religious ceremonies."

Northwest of the Himalayas in Western Tibet there is a large region with over a thousand caves. They are referred to as the "Sacred Caves," "Mystery Caves," and the "Shangri-La Caves." Similarly, there is a stone with three openings in the small area in front of Cemetery House. The stone is located northwest of the Dove House (the Himalayas replica), just as we find on contemporary maps. Presumably, the three holes in the Cemetery House stone also represent the three paths in Buddhism.

CEMETERY HOUSE, UXMAL.

LEFT AND CENTER: SKULLS AND BONES AT CEMETERY HOUSE COURTYARD, UXMAL. RIGHT: TIBETAN SKULL. IMAGE BY AS23.

LEFT: REPLICA OF SACRED TIBETAN CAVES AT CEMETERY HOUSE COURTYARD, UXMAL. RIGHT: SACRED CAVES OF NGARI, TIBET-CHINA. IMAGE BY UDOMPETER.

Chapter 12

Russia, China, Cambodia, Thailand and Sri Lanka

Russia, Mongolia, and the Great Wall

The "North Group" of temples represents Mongolia and Russia, in Asia's far north. However, archaeologists have said little about this set of structures. A wall surrounding most of Uxmal includes the Himalayas replicas. Therefore, it does not appear to represent the Great Wall of China. There are several other candidates for the Great Wall replica in Uxmal. However, none are outstanding.

China

The "Pyramid of the Fortuneteller" (Adivino) represents Beijing (ancient Peking), China. An estimated one-third of China's population believes in fortunetelling. This structure has another name however—"Pyramid of the Magician," which describes China's "Fangshi" practitioners (alchemists, astrologers, magicians, etc.).

HOUSE OF THE FORTUNETELLER, UXMAL (2015). PHOTO BY AUTHOR.

HOUSE OF THE FORTUNE TELLER SEEN FROM THE GOVERNOR'S PALACE.

China: Tiananmen Square

Uxmal's "Courtyard of the Birds" represents Beijing's Tiananmen Square. Both "Tiananmen" and the Mayan "birds" symbolize heaven. Another shared feature of Tiananmen Square and the Courtyard of the Birds is a central pillar. In each case it is aligned with a large gate and centered with pillars on the perimeter. The "House of the Iguana," may represent iguanas, as they are a Chinese delicacy.

TIANNANMEN SQUARE WITH A CENTRAL MONUMENT TO THE PEOPLES' HEROES, PHOTO BY KEVINAI.

COURTYARD OF THE BIRDS, WITH CENTRAL MONUMENT ON A SMALL, RAISED, SQUARE, UXMAL.

China: Buddhist Monastery

Immediately west of the Courtyard of the Birds is "The Nunnery," which represents a Buddhist monastery near Tiananmen Square. Uxmal's nunnery once included a replica of Buddha in his traditional, seated position with legs folded. Presumably, the architect gave Buddha a suffering facial expression (see image below) because suffering is an important concept in Buddhism. He seems to have a bell in his mouth, which may indicate chanting. The Nunnery's statue is different from the typical stoic or smiling Buddha.

LEFT: SITTING FIGURE, THE NUNNERY, UXMAL,
PHOTO BY MACDUFF EVERTON.
RIGHT: BUDDHA, IMAGE BY: FLORIAN BLUMM.

China: The Taijitu Symbol

The west side of the Pyramid of the Fortuneteller has roughly two dozen taijitu symbols. In Asia they mean "the great ultimate," and were the precursor to the yin yang symbol.

LEFT: THE "TAIJITU" SYMBOL AT THE PYRAMID OF THE FORTUNETELLER, UXMAL. CENTER: THE YIN YANG SYMBOL. RIGHT: TAIJITU SYMBOL.

China: Qinghai

Qinghai, China is northeast of Lhasa. The image below presents a view (through the trees) of Qinghai from Lhasa on the map. Since Qinghai and Lhasa have roughly the same altitudes in actuality, the elevated view in the image below reflects only the extra elevation at Potala Palace.

The ball court represents the fault line on the Tibetan Plateau. It is formed by tension between the Eurasia (tectonic) Plate to the north, and the India Plate to the south. The area of Yushu in Qinghai experienced a major earthquake in 2010. In the ball court image below, the observer is in Qinghai, and looking south toward the Palace of the Dalai Lama (Potala Palace) on a hilltop in Lhasa, Tibet (China).

A VIEW OF THE QINGHAI REGION REPLICA FROM LHASA ON THE MAYAN MAP AT UXMAL.

ONE OF SEVERAL TEMPLES REPRESENTING QINGHAI, CHINA.

LOWER: BALL COURT REPLICA OF THE QINGHAI, CHINA EARTHQUAKE ZONE. UPPER: REPLICA OF POTALA PALACE (LHASA, TIBET).

Cambodia

Uxmal's "Pyramid of the Old Woman" refers to Angkor's (ancient Cambodia's) Temple of Banteay Srei. It predates Angkor Wat by two centuries. The "old woman" refers to Karaikkal Ammaiyar, the elderly female deity whose image is carved at the temple. She was a very devout human who was subsequently deified.

BANTEAY SHREI, ANGKOR. IMAGE BY ADRIAN WOJCIK.

HOUSE OF THE OLD WOMAN, UXMAL.

Thailand

Uxmal's House of the Phalli contains anatomically correct statues similar to those at the modern day Phallic Temple in Thailand. The Maya structure represents this temple and Thailand's role in modern sex tourism.

Sri Lanka

On the Mayan map, Sri Lanka is called the "House of the Cedar Tree." In actuality, Sri Lanka has many cedar trees. However, a special connection between the tree and the island is not apparent.

The North and South Poles

Chapter 13

The North Pole and Arctic Region

The Mayan map presents the Arctic Region north of a stone wall that represents the Arctic Circle. At the northernmost point in this area there is a circle of water. It represents the water at the top of the world, and it is called the "Sacred Cenote." A cenote is a natural lake in a sinkhole that is usually a few dozen meters deep. The architect rounded the walls of the Sacred Cenote to represent the watery circle of latitude at the North Pole region.

Inuit Senicide

Until roughly 1939 the Inuit tribes of the far north practiced senicide—killing the elderly who couldn't keep up with the group. They would toss them into icy water, or leave them on the ice to perish from exposure. Those who were strong enough to return to the tribe were welcomed back.

The Mayas had a similar practice at the Sacred Cenote, the North Pole region. They would throw people into the waters, who, in turn, would offer prophetic messages before drowning. As was the practice among the Inuit, Maya victims who managed to climb out of the Sacred Cenote were welcomed back to Mayan society. One such victim was Hunac Ceel, who offered a prophetic message. He climbed out of the cenote and became a respected ruler of Mayapan. Thus, according to the map theory, the people who were thrown into the cenote—however willing or unwilling—were not victims of "sacrifice," but of a deadly pageant. In other words, they were not killed to appease angry gods, but to present the story of the Inuit in the Hall of Records.

Inuit Prophecy

In recent years, Inuit Elders have offered prophetic messages about earth's climate. They say our planet has "wobbled," which has caused the sun to rise in different locations than previously, which has caused the morning sky to heat up more quickly. Thus, the prophecies of the Sacred Cenote victims may refer to Inuit Elder prophecy.

The Uniqueness of the Sacred Cenote

Evidence of the uniqueness of "sacrifice" at the Sacred Cenote is that, among an estimated 30,000 cenotes across the 4,400 Maya cities (reported by Brown and Witschey in 2011), sacrifices were only made at one cenote in one city. In 2014 Bradley Russell's archaeological team found human remains in a cenote near Mayapan. However, they found no evidence of human sacrifice in it.

Arctic Gold

Another outstanding feature of Earth's far north was the Gold Rush of the late 1800's. Stampeders sailed into Dyea, climbed over the dreaded Chilkoot Pass, and panned for gold along the Klondike River in the Yukon Territory. The Gold Rush is symbolized by roughly seven million dollars (at 2014 prices) of valuables—including several golden monkeys—at the bottom of the Sacred Cenote. One of the monkeys is actually shaped like a monkey trap, which operates as follows. When a monkey inserts his hand into an unmovable object to retrieve food, his closed fist cannot be extracted. The reason is that the inside of the object is wider than its opening. Therefore, the trap only works on monkeys that are too "greedy" to release the food before the hunter arrives. Accordingly the golden monkeys at the Sacred Cenote represent gold, and the type of greed known as "gold fever."

GOLD SEEKERS TRAVELING TO NOME, ALASKA.

The Blue Waters of the North

The Mayas poured 2,500 gallons of "Maya Blue" paint into the Sacred Cenote. Some believe the resulting 14 ft. layer of paint at the bottom of the cenote came from sacrifice victims. However, that is unlikely since it would amount to about 25 gallons per victim. A more plausible hypothesis is that the Mayas (the Architect) painted (dyed) the cenote to make it resemble the azure waters of the North Pole region.

SACRED CENOTE, CHICHEN ITZA.

The Arctic Cordillera and Arctic Circle

The semi-circular structure of stones overlooking the Sacred Cenote represents the Artic Cordillera, sparsely populated by Inuit Native Americans. On the Mayan map their society in Northern Greenland is represented by the tiny, Northern Temple of Xtoloc. The temple consists of a main section and a smaller section that runs along the south rim of the cenote. The images of people carved on the temple walls represent the Inuit who live along the coast of Baffin Bay near the North Pole. The map presents the Arctic Cordillera (mountains) at a higher elevation than the temple to symbolize Inuit life at the lower elevations.

On the Mayan map the Arctic Circle is indicated by a massive, straight, stone wall. It is farther north than it would be on a map drawn to scale. However, it is accurately north of the Iceland temple and south of the Arctic Cordillera.

NORTHERN TEMPLE OF XTOLOC AND SEMI-CIRCLE.

THE ARCTIC CORDILLERA.

LOOKING WEST ALONG THE "ARCTIC CIRCLE" AT CHICHEN ITZA.

Chapter 14

The South Pole and the World Tree

The South Pole

Roughly two miles south of the Sacred Cenote at Chichen Itza's southernmost point lies the Cave at Balankanche. It was designed so that surface rainwater encircles a platform and creates an island. Since Balankanche represents the extreme south, the island is Antarctica—the South Pole.

The World Tree

The Antarctica replica is home of the Mayan "World Tree" (called the "Wakah-Chan"), which is actually two trees close together, and made of stalactites and stalagmites. Maya scholars (e.g. Linda Schele) indicate that the tree "...is the central axis of the world," which is consistent with the South Pole hypothesis.

The Mayan Map's "Garden of Eden"

Over the centuries the Hall of Records has been described as a depository of important information. Accordingly, the Antarctica replica contains a replica of the "Garden of Eden" (i.e. the Bible, the Legend of Gilgamesh, etc.). Both gardens:

- were sacred spaces where ancient people prayed.
- contained two trees side-by-side.
- had only one tree that was touched by human hands.

- had a river that flowed into it, and then split into four rivers.
- had access sealed off.
- were prophesied to be rediscovered.
- were an axis mundi (where gods and men meet).
- had an intelligent serpent who lived in the tree.

Sealing the Cave at Balankanche

The tree at Balankanche had been sealed off for perhaps thousands of years before a Maya named Jose Gomez discovered it in 1959. According to researcher Stanislav Chladek, immediately after the tree's discovery, Maya priests held rituals for 24 hours to appease "the balam"—guardian spirits. These would represent the guardian "angels" who guarded the tree in Genesis. The Maya ritual culminated with the entrance to the tree being symbolically re-sealed for two days before it was reopened for visitors. Finally, both gardens are holy sites, evidenced by ancient prayer artifacts at the Balankanche garden.

The Vision Serpent

Maya mythology indicates that a "Vision Serpent" resides among the branches of the World Tree. He walks upright and can communicate with humans. Thus, he is similar to the serpent in the Garden of Eden, who is associated with "the fall" of man. "Vision" is consistent with the Biblical serpent's statement that "your eyes will be opened." However, it isn't clear whether the Mayas regarded this serpent as good or evil. Below we see the Vision Serpent rising out of a fire and walking upright.

MAYA VISION SERPENT, PHOTO BY PRISMA ARCHIVO.

THE WORLD TREE AT BALANKANCHE,
CHICHEN ITZA PHOTO BY eXpose.

Aurora Australis

If the Garden of Eden was in Antarctica, then the "Angel with the flaming sword" may have referred to the Aurora Australis—the southern version of the Aurora Borealis. Researcher Robert Argod has theorized that the Garden of Eden is in Antarctica, and that ancient Polynesian myths and shifting lithospheric plates support the theory.

The Western Hemisphere: Europe

Chapter 15

Italy: Pillars and Contours

Chichen Itza has a great many pillars on the east side of the Grand Plaza (the Atlantic Ocean replica). However, it has few pillars on the west. The European side pillars indicate the Roman Empire, and the American side pillars indicate Roman influence.

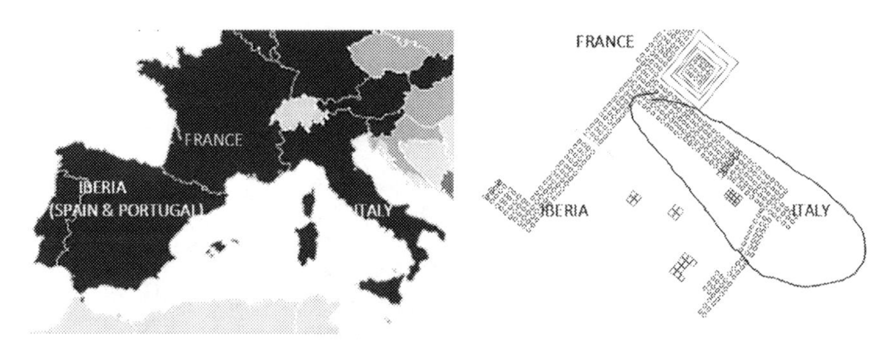

LEFT: MODERN MAP (SEE IBERIA AND ITALY).
RIGHT: MAYAN MAP.

Italy's Unique Geography

Italy is shaped like a boot with a heel, and so is its Mayan replica. Similarly, the map's small stone structures represent the Mediterranean Sea Islands. The map's Italian coastline has been smoothed, and thus omits the finer details of its natural counterpart.

On the Mayan map the ball court on the east side of Italy represents the Adriatic (tectonic) Plate, which surrounds the Adriatic Sea. The ball court beneath the "foot" of Italy represents the Hellenic Trench, where devastating earthquakes have struck the island of Crete at different points in history.

Chapter 16

Italy: Tunnels and Waterways

One of Italy's most outstanding features is its ancient underground infrastructure, including hundreds of miles of catacombs and aqueducts. Some categorize the aqueducts as one of the Seven Wonders of Ancient Rome.

ROMAN CATACOMBS. PHOTO BY VLADIMIR KOROSTYSHEVSKIY.

TUNNEL AT THE "ITALY" TEMPLE, CHICHEN ITZA.

ROMAN AQUEDUCT, PHOTO BY PIPPAWEST.

REPLICA OF ROMAN VIADUCT,
SICILY REPLICA, CHICHEN ITZA.

Chapter 17

Italy: Tic Tac Toe

Perhaps the most enigmatic structure at Chichen Itza is a set of twin Tic Tac Toe boards near the Sicilian replica. Each board is carved in stone, and each game has the same strange outcome. The x's went six times, the o's went three times, and the o's won. These cannot represent actual games because the outcome is impossible. Moreover, since archaeologists haven't found any Mayan Tic Tac Toe boards with moveable pieces, it is unlikely that this particular structure is a monument to a Mayan game. It is more likely that the game is an encryption—to protect the map. As a reminder, the interpretive heuristic for the Yucatan Hall of Records is that everything (myths, rituals, symbols) indicates a specific location (e.g. the specific country); or it is interpreted within the context of one.

TIC TAC TOE BOARD AT CHICHEN ITZA.
IMAGE BY FRONTPAGE.

Consider the following. The game boards are in the Roman Empire section of the Mayan map, where Tic Tac Toe originated, and where x = 10 in Roman numerals. When one of the board game matrices is read from left to right and top to bottom—which was how items were read in the Roman Empire—the result is 10100 10010 01010. Those numbers had no apparent meaning to the ancient Romans. However, in modern times they represent "2, 3, 4," and "9, 8, 5" in a "Two-out-of-Five" binary code sequence. Computer scientists developed it to accommodate decimals in telecommunications. Each trio of numbers will be discussed in turn.

Tropical Landmarks

The numbers "2, 3, 4" are significant to both the Roman Empire and to cartographers. The earth rotates at an angle of 23.4 degrees—which defines the Tropics of Cancer and Capricorn (at 23.4 degrees)—on either side of the equator. Thus, on the Mayan map at Chichen Itza, the northern Tic Tac Toe board indicates the Tropic of Cancer, and the southern board indicates the Tropic of Capricorn. The middle symbol—which seems to include a Mayan fire symbol (for the sun)—represents the equator.

Including the tropics on the Mayan map is consistent with the architect-cartographer's inclusion of the earth's poles and the Arctic Circle. He could have built the tropical and equatorial landmarks as long stone walls running east-west across Chichen Itza. However, since the map is not drawn to scale, the walls would have placed certain nations too far north or south. The Tropic of Cancer's relevant significance is that it is where the Roman Empire stopped its southern advance, because the Sahara Desert was too hot.

TIC TAC TOE BOARDS. PHOTO BY AUTHOR.

Temporal Landmarks

As stated above, the 10100 10010 01010 binary sequence can also be read as "9, 8, 5." That number is also significant to the Roman Empire, which began in 509 BC and ended in 476 AD—a period of 985 years. Thus, the game board structure has both geographic and historic meaning.

Chapter 18

Italy: Pantelleria Vecchia Bank

Pantelleria Vecchia Bank ("PVB") is currently at the bottom of the Mediterranean Sea between Sicily and Tunisia in North Africa. It sank 10,000 years ago, and should not be confused with the nearby island, "Pantelleria Vecchia." On the Mayan map, PVB is called the "Temple of the Little Tables" because its structures are a smaller version of the ones at the "Temple of the Big Tables," which represent England's Stonehenge. In 2015, archaeologists Emanuele Lodolo and Zvi Ben-Avraham from the University of Tel Aviv discovered what many archaeologists refer to as a "Stonehenge" on PVB. For scholarly reasons, however, their *Journal of Archaeological Science* article does not use the term "Stonehenge." Rather, it uses only scientific terms (e.g. "megaliths").

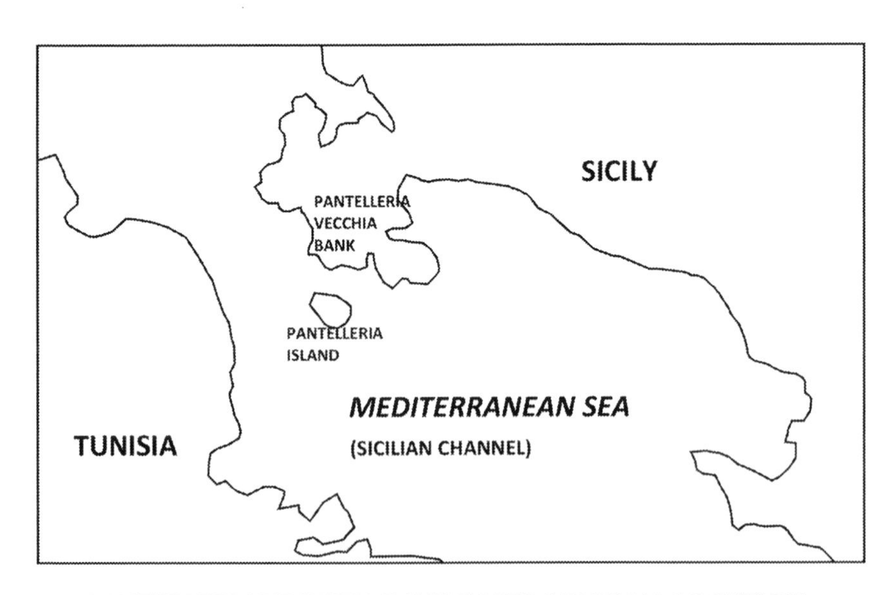

PANTELLERIA VECCHIA BANK IN ITS ORIGINAL LOCATION.

WESTERN EDGE OF THE SICILY REPLICA AND NORTHEAST SECTION OF THE TEMPLE OF THE LITTLE TABLES (PANTELLERIA VECCHIA BANK REPLICA), RECTANGLE INDICATES SPACE BETWEEN THE TWO ISLANDS, PHOTO BY AUTHOR.

ATLANTS FROM THE "TEMPLE OF THE LITTLE TABLES" (1800'S). CORNELL UNIVERSITY.

REPLICA OF A MEDITERRANEAN SEA
ISLAND, CHICHEN ITZA.

Chapter 19

France: Divinity

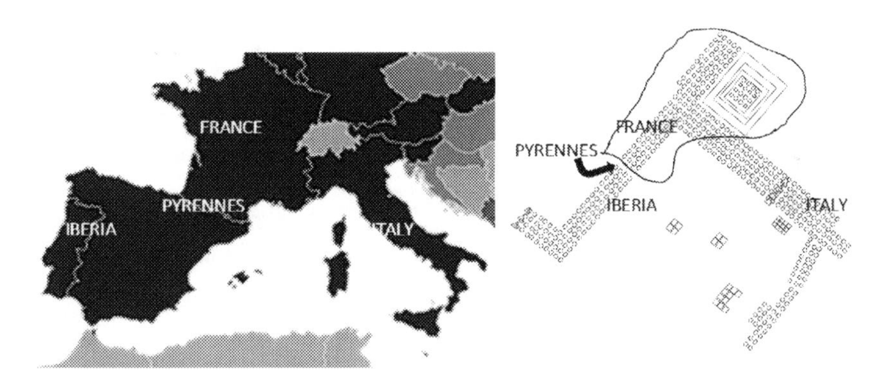

LEFT: FRANCE AND PYRENNES. RIGHT: MAYAN MAP.

The Cathedral of Notre Dame

The twin sections of the "Temple of the Warriors" at Chichen Itza fit the general description of the Cathedral of Notre Dame in Paris. The name of the Mayan temple refers to its images of soldiers with backpacks, sticks that produce smoke at one end, a squirrel eating a human heart, etc. As will be shown in the "Washington, D.C." portion of this book, an animal eating a human heart refers to a ruler. Hence, the squirrel refers to French Leadership.

The Merovingian

The "God Impersonator" seated on a jaguar and holding a scepter at the Notre Dame replica represents Charlemagne on a horse, holding a scepter. "God Impersonator" may refer to the Merovingian family's claim to divinity. During the 1980's, a Frenchman named Pierre

Plantard de St. Clair attempted to raise international support for what he claimed was his right to rule France—as the last in the holy bloodline of Merovingian kings. His altered story became part of the plot to *The Da Vinci Code*.

TEMPLE OF THE WARRIORS, CHICHEN ITZA.

Chapter 20

France: World War II

The Temple of the Warriors describes the World War II Nazi invasion of France. Its hieroglyphics refer to a place on a sea coast (France) that is attacked by soldiers—in dark gray uniforms—from the north (Germans). It is clear that the water is an ocean, given its aquatic life. During the fight (World War II), citizens are described as "screaming and fleeing" with large bundles on their backs. A large, white, bird is shown diving like a jet fighter, and people and houses are on fire. One particular ethnic group is taken away naked, as captives. In one of the images an interracial couple flees as the mother carries a toddler. The father is wearing a backpack and a World War I French Army style helmet. He holds a walking stick which is shoulder height and with the gnarly imperfections of wood. Some images show men wearing replicas of modern army helmets and carrying backpacks and short "walking sticks" (according to anthropologists). These sticks are perfectly straight, held in the center, and produce fire at one end. Thus, they appear to represent rifles.

The scene also describes soldiers crossing the ocean from the west in "canoes" (U.S. Navy vessels). It also says, "Warriors sail in canoes past a 'Maya' coastal community with a 'Toltec' temple, and there is no apparent confrontation." This may refer to Armistice Day—the end of the war. Some of the temple inscriptions and images are found in *Merchants, Markets, and Exchange in the Pre-Columbian World* by Hirth and Pillsbury.

Normandy and the Atlantikwall

There are many temples at Chichen Itza. However, the only one with a large stone slab in front of it—in the northwest corner—is the Temple of the Warriors. The slab represents the beach at Normandy, site of the American World War II D-Day landing. The tiny "windows" overlooking the slab on the north side of the temple represent Germany's "Atlantikwall," which consisted of fortified gun bunkers along the French coast. Note: In the "France" mural, the attacking "German" soldiers are wearing dark gray uniforms. The captives are nude—hands tied behind their backs. The attackers and their captives are white—not brown like Mayas.

RIGHT: TEMPLE OF THE WARRIORS "FRANCE." LOWER CENTER: "NORMANDY." CENTER: ENGLISH CHANNEL & TUNNEL.

A Fire-breathing Dragon

The Notre Dame replica at the Temple of the Warriors shows a man walking beside a fire-breathing dragon as another man retreats. This could refer to a literal fire-breathing dragon (for some reason) or to weaponized fire (e.g. the massacre at Oradour-sur-Glane, flamethrowers, etc.).

Chapter 21

France: Chac Mool and Landmarks

The gap between the pillars south of the Temple of the Warriors represents Roland's Breach in the Pyrenees. These mountains cross the width of the peninsula and form the border between France and Iberia (Spain and Portugal).

The Pyrenees are aligned with the southwest corner of Corsica in the region of Filitosa and Ajaccio. Filitosa is known for its ancient "Soul Stones" (grave protectors), and Ajaccio is the birthplace of Napoleon Bonaparte. Accordingly, a line traced from the replica of Roland's Breech leads to a lone pillar with a soldier carved on it. It is the only stand-alone pillar at Chichen Itza, and represents the birthplace of Napoleon Bonaparte. He is facing the French mainland.

Representing the era of French colonialism, France's Chac Mool looks toward the Americas from a replica of Notre Dame. The architect created the replicas a thousand years before the "originals" were built.

NAPOLEON'S "MENHIR," CHICHEN ITZA.

ROLAND'S BREACH REPLICA ON THE MAYAN MAP.
IMAGE BY AUTHOR.

LA BRECHE DE ROLAND, PYRENEES, FRANCE.
IMAGE BY ROTATEBOT.

Chapter 22

France: Blue Skin

The Fugates

According to ABC News (Feb. 2012), in 1975 a newborn baby was rushed to the University of Kentucky Medical Center because his skin was blue. However, an investigation revealed that he was a descendant of a French orphan named Martin Fugate, who had come to America during the early 1800's (c.a. 1820). His lineage became known as the "Blue People of Kentucky," and as the "Blue People of Troublesome Creek."

Fugate had a genetic trait that resulted in "methemoglobinemia," a disease that made his skin, and that of his descendants, blue. Since this was genetic, it means Fugate inherited it from his ancestors in France. However, there is no evidence that—as presented among the Mayans—the Fugates' hearts were removed. Therefore, the Fugates are not likely to have inspired the Maya ritual of painting naked captives blue before removing their hearts.

Gaelic Invaders

A better explanation for the Mayan ritual of painting sacrifice victims blue—and the one proposed here—is that it reenacted (or pre-enacted) the Gaelic attacks against France. Centuries before the reign of Robert the Bruce, Briton's Gaelic warriors would remove their clothes and paint themselves woad-blue prior to battle. There is historical evidence that they attacked France, and that the earliest use of blue woad paint was in France. Thus, the Mayan ritual of painting naked enemies blue before killing them at the France temple (Temple of the Warriors) was historically (or, futuristically) accurate.

Robert the Bruce

Removing the victims' hearts and displaying them (the Mayan priest held each heart up high) is consistent with the story of the Gaelic leader, Robert the Bruce, who lived a few centuries after the time of the blue woad Gaelic attackers. However, his will specified that, upon his death, his heart be removed and carried into future battles. Presumably, it was held high as a source of inspiration. In other words, the Mayan priests removed blue-painted prisoners' hearts to tell the composite story of the Gaelic invaders and Robert the Bruce.

Chapter 23

Iberia

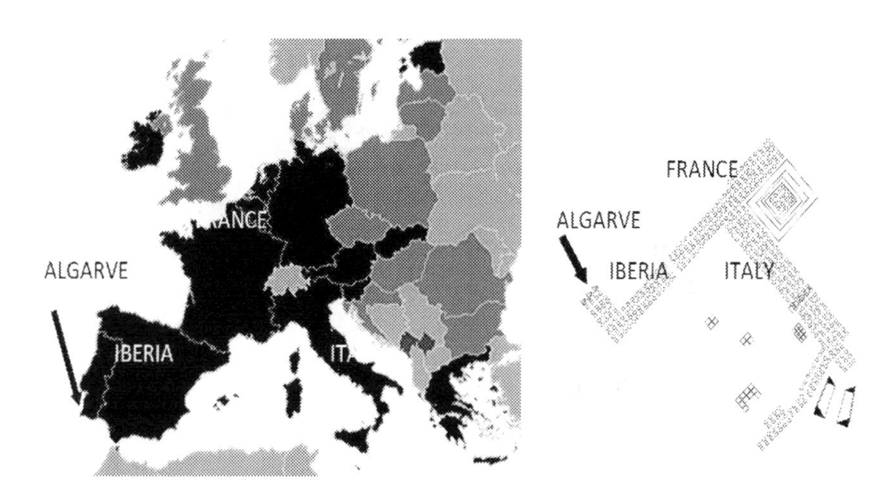

LEFT: MODERN MAP OF IBERIA.
RIGHT: MAYAN MAP OF IBERIA.

SEA CAVES AT THE ALGARVE, FARO-PORTUGAL.
PHOTO BY GERGO KAZSIMER. DREAMSTIME.

The backwards "L" on the Mayan map represents the eastern and southern coasts of Iberia. The architect used sea caves at the Algarve as the primary landmark here. While there are caves all over the world, the architect specified three, equidistant, A– framed caves that open toward the Atlantic Ocean.

The Algarve is one of the possible locations of the lost continent of Atlantis. Dr. Roger Coghill believes the Algarve matches Plato's description of Atlantis, even to the point of its black, yellow, and red colors. According to Coghill, the word "Atland" was found carved into some of the ancient structures at Faro/Algarve. Researcher Peter Daughtrey also places Atlantis in the Algarve.

The Algarve vs Gorman's Cave

The Algarve and the sea caves at Gibraltar are similar in appearance and are located in the same general vicinity. Gorman's Cave is the most notable one at Gibraltar, as archaeologists describe it as one of the last know habitations of the European Neanderthals. However, given their location and number (three at the Algarve versus four at Gibraltar), the temple at Chichen Itza better replicates the Algarve.

THE ALGARVE SEA CAVES REPLICA ON THE MAYAN MAP.

THE ALGARVE SEA CAVES REPLICA ON THE MAYAN MAP.

Chapter 24

England, Scotland and the English Channel

Stonehenge

The open area northwest of the France temple represents the English Channel. It contains a small stone bridge (English Channel Tunnel replica) that connects it with the England temple ("Temple of the Big Tables"). The Temple of the Big Tables represents Stonehenge, an ancient megalithic site in England.

The British Royal Family

The Temple of the Big Tables also shows lions against a red, yellow and blue background. This represents the British Royal family's coat of arms, which typically includes lions against a red, yellow and blue-colored background on a shield. The lions at the Chichen Itza temple are shown approaching and leaving groups of castle towers. This may represent the changing monarchs.

BRITISH ROYAL FAMILY SYMBOLS FROM
THE TEMPLE OF THE BIG TABLES.

STONEHENGE, 1920.

TEMPLE OF THE BIG TABLES, USED WITH PERMISSION-CARNEGIE INSTITUTION.

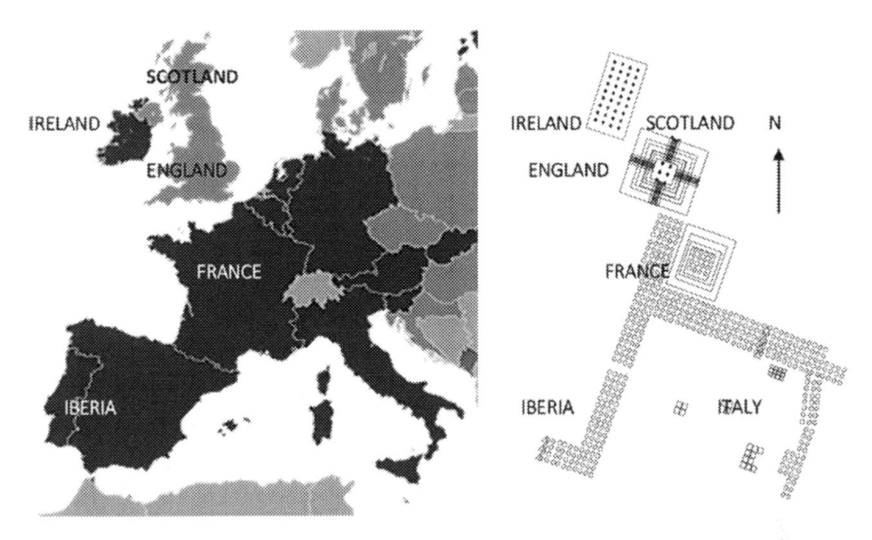

MODERN AND MAYAN MAPS OF ENGLAND.

LEFT: ENGLISH CHANNEL TUNNEL.
RIGHT: MAYAN REPLICA OF ENGLISH CHANNEL TUNNEL.

Chapter 25

Ireland and St. George's Channel

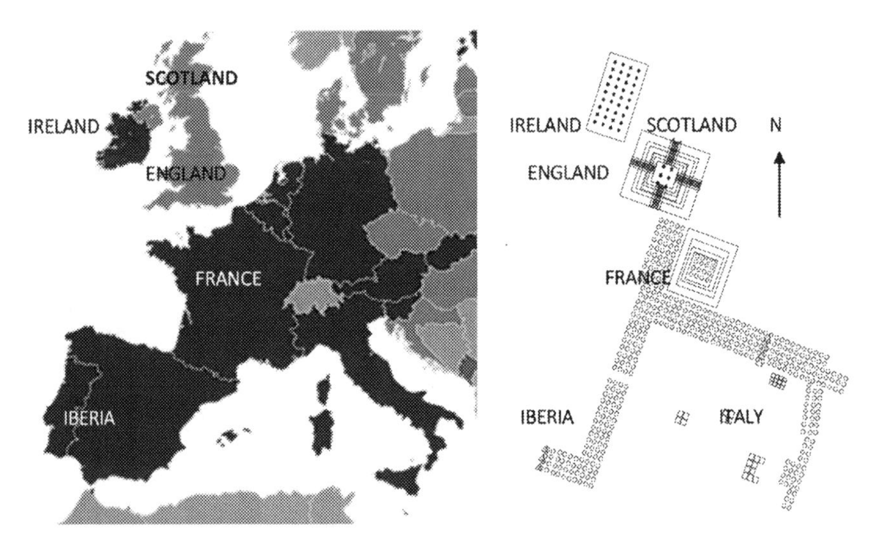

LEFT: MODERN MAP. RIIGHT: MAYAN MAP.

North of England on the Mayan map is a narrow space which represents St. George's Channel. It separates Scotland and England from Ireland. There is no connecting structure between the two landmasses on the Mayan map because there are none in actuality. The Ireland temple is filled with pillars, perhaps to indicate the "Giant's Causeway." The actual causeway consists of roughly 40,000 basalt pillars from an ancient volcanic eruption. They are naturally carved into hexagons and stand roughly 100 meters tall, although some only rise a few inches aboveground.

LEFT: PILLARS REPRESENTING GIANTS CAUSEWAY. OVAL REPRESENTS THE CELTIC SEA AND ST. GEORGE'S CHANNEL.

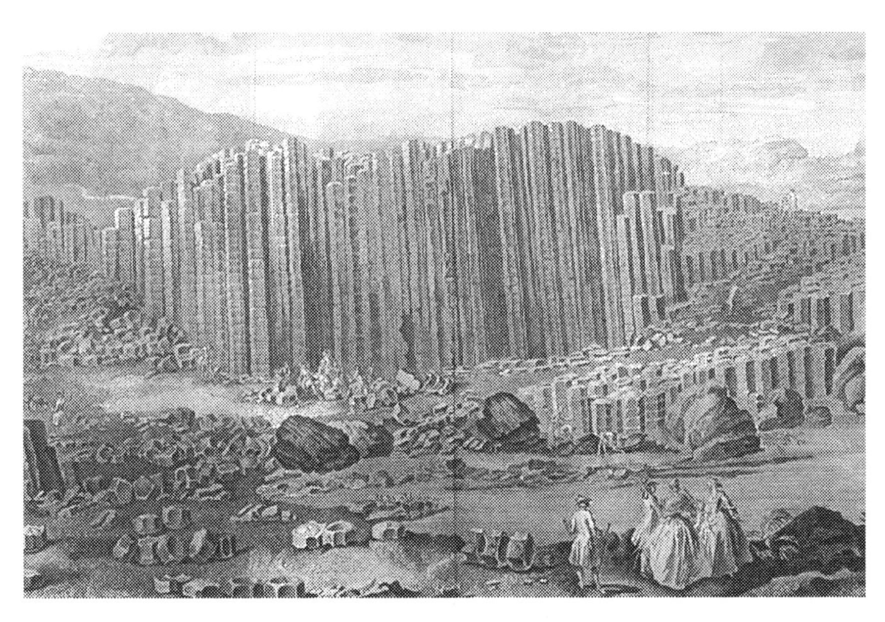

ENGRAVING OF SUSUANNA DRURY'S "A VIEW OF THE GIANT'S CAUSEWAY: EAST PROSPECT" (1768).

Africa

Chapter 26

Morocco and Libya

Morocco

On the Mayan map at Chichen Itza, Morocco is represented by a ball court. It symbolizes the South Atlas earth quake zone on the African Fault line, the scene of the 1960 Agadir earthquake. There were many thousands of casualties. Currently, the ball court is unexcavated, as the following image indicates.

BALL COURT IN "MOROCCO," CHICHEN ITZA.

Libya

The Mayan map includes two landmarks for Libya. The first is an arch about 15 ft. tall and north of a stone wall. The arch represents the 130 ft. tall Forzhaga Arch at Tadrart Acacus in Southern Libya, near the Algerian border. It cannot represent the natural arches in nearby Chad because they are *south* of the Yellow Nile replica (western branch), which dried up in 10,000 B.C. Today, the Yellow Nile is called the "Wadi Howar." A "wadi" is a dry river bed. The Yellow Nile river bed was recently discovered via satellite technology. Forzhaga Arch is outstanding because of its size.

The Mayan arch at Chichen Itza, therefore, is not a glorious welcome sign. In the image below called "Stone Wall Reconstruction," one can see yellow tape from recent reconstruction. The well-meaning effect was to present the arch as a type of Arc de Triomphe in Paris. However, the historic and prophetic messages in the Hall of Records will be more difficult to interpret as its structures are altered.

The second Libyan landmark on the Mayan map is a temple called "The Market." It represents the city of Leptis Magna. National Geographic described Leptis Magna as a city "built around the harbor." Because of its excellent natural port, Leptis Magna was the largest market on the North African coast during the Roman Empire.

FORZHAGA ARCH AT TADRART AKAKUS, LIBYA – PHOTO BY LUCAG.

STONE ARCH AT CHICHEN ITZA (FORZHAGA ARCH REPLICA).

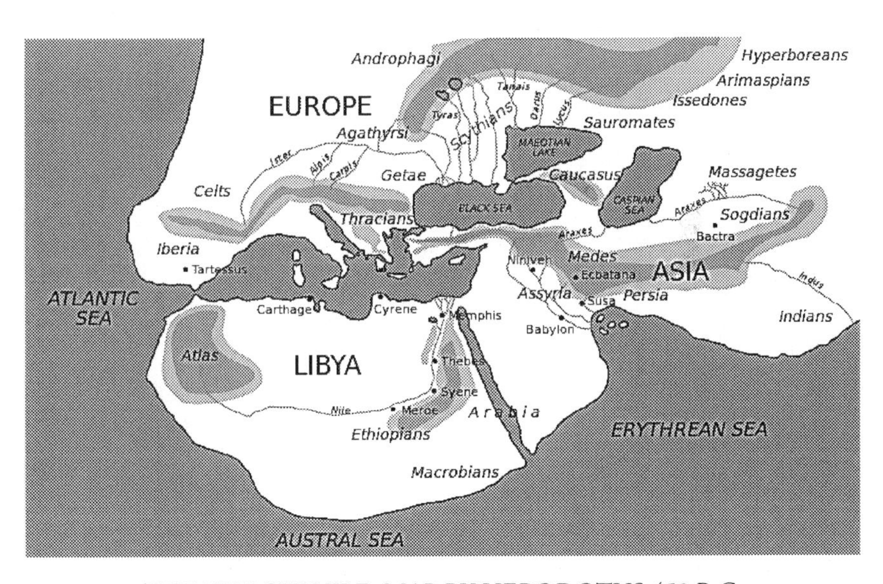

THE YELLOW NILE, MAP BY HERODOTUS 450 B.C.

STONE WALL (YELLOW NILE), CHICHEN ITZA.

STONE WALL (YELLOW NILE) PASSING THE ARCH (FORZHAGA ARCH REPLICA) AT CHICHEN ITZA.

STONE WALL RECONSTRUCTION OF THE
YELLOW NILE REPLICA IN 2015.

Chapter 27

Chad

The Yellow Nile terminated near Lake Chad in the African nation of Chad. If the map theory is correct, then, for some reason, the temples of Xtoloc are located in places of extreme weather. The northern temple of Xtoloc is located at what represents the North Pole, while the southern one represents the center of the African Sahel—known as one of the hottest and driest places on earth. Droughts there can last for centuries. Perhaps Xtoloc indicates regions with little precipitation, temperature aside.

CENOTE OF XTOLOC, CHICHEN ITZA.

LAKE CHAD, AFRICA, PHOTO BY WALTER MITTELHOLZER.

Chapter 28

Egypt and Sub-Saharan Africa

Egypt

On the Mayan map, Egypt is the colonnade east of the "Market" (Libya) and west of the "Steam Bath" (Israel). Presumably the architect did not use pyramids to identify Egypt because, since there are pyramids all over Chichen Itza, pyramids would not distinguish Egypt from any other place. However, one of Egypt's distinguishing natural features is the Nile River. Presumably, the Nile is more sacred and revered than the pyramids, since it predated them.

On the Mayan map the Egyptian part of the Nile is represented by the north-south section of the stone wall (mentioned previously) that forks east and west. The forks converge and then turn west to become the Yellow Nile, as shown previously on the map by Herodotus in 450 B.C. The Yellow Nile once ran west for roughly 1,000 miles to Lake Chad, which the architect represented with the cenote of Xtoloc (South). Thus, the Mayan map of the Nile, and Herodotus' map of the Nile, are very similar.

Archaeologists' maps of Chichen Itza rarely include this wall or its fork. However, the wall's gentle curve north and its fork are observable within a five-minute bushwhack of the Mayaland the hotel.

THE STONE WALL FORK (EAST AND WEST NILE).

The Great Pyramid of Giza

According to Egyptian mythology, long ago, a small "mound,"— accompanied by the invisible god of time—fell from the sky into the waters at Heliopolis (a suburb of Cairo). The time god (who, presumably, can travel through time) was carrying a round object (with feathers on it). The mound sank into the water, and then resurfaced to become the foundation of the world.

I interpret this to mean that a space capsule fell into the Nile and buoyed back up. The time god—the Bennu bird—who glides like a crane, represents the space capsule's diagonal approach to the water. The round object with feathers (representing flight) represents the capsule's parachute.

The myth also indicates that the mound was replicated as the "Benben Stone," which subsequently became the tops of obelisks. The time god indicates time travel, and the Benben became the prototype for the Egyptian pyramids.

The original Benben Stone is in the Cairo Museum. Carved on it is the face of a gorilla. Presumably, he represents a "test-gorilla." He is shouting a message about Amenemhet (the 3rd) seeing the "Lord of the Horizon as he sails in the sky." Two of the first hieroglyphics from his mouth are the ones that precede the images of modern air and water craft at the Temple of Seti I at Abydos.

The space gorilla was subsequently honored in Egyptian mythology as Thoth, the god of wisdom, magic, and learning. The Great Pyramid, then, is a monument to Thoth, who is either the space gorilla, itself, or the brains behind him.

The following evidence is based on my experience at the Great Pyramid (summer solstice, 2010). To reach the "King's Chamber," one must walk—bent over like an ape—up a low shaft. Then one walks upright, and then crawls—ape movements. The King's Chamber houses one object—a massive, stone "sarcophagus" (a coffin). It has no lid because it is not a sarcophagus, but rather a "suitcase;" a replica of the chair-beds that America's monkey astronauts used during space flight.

Inside the King's Chamber I heard what I initially assumed was a 747 airplane passing overhead. However, while the plane approached, it never "flew away." The sound remained around the pyramid. Subsequently, some New Age experts in Giza told me they were familiar with the sound, and that the acoustics are better from inside the sarcophagus.

The Grand Gallery may indicate the rocket's stages, and the three levels of the pyramid may indicate the three levels inside a space capsule.

My overall interpretation is that there was once a reenactment ritual at the Great Pyramid, which entailed walking like an ape (Thoth)—up to the King's Chamber. There, one would lay in the sarcophagus/suitcase, hear the speeding "rocket," and meditate on Thoth and the Bennu Bird—the "gods" of *wisdom* and *time travel.*

The "space monkey" hypothesis is plausible in that apes don't regularly fall from the sky in tiny pyramids that sink into the sea and then pop up again. Moreover, it seems beyond chance that obelisks resemble space rockets, and that astronauts ride in the "Benben" part—the capsules.

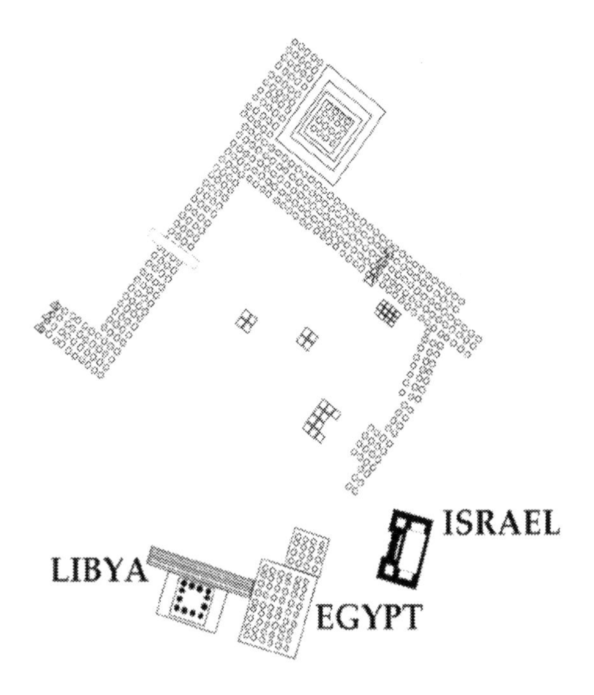

MAP OF MIDDLE EAST AT CHICHEN ITZA.

Sub-Saharan Africa

The Mayaland Hotel rests upon what were once temples that represented Sub-Saharan Africa. A few foundations of small to medium-sized temples remain. During construction of the hotel, workers found an earthen jar containing the remains of a Mayan infant. Archaeologists Eduardo Heredia, Gabriel Canul, Francisco Ruiz, Jose Osorio, and Jose Arias compared the artifact with others at Chichen Itza and concluded that:

> "…in the case of Chichen Itza, we could be facing ritual activities of infant sacrifices…"

Since the hotel area is south of the stone wall that represents the Yellow Nile, the infant graveyard (assuming there were other jars) represents the eastern portion of Sub-Saharan Africa, where ritualized child sacrifice and other forms of infanticide are still practiced. Thus, the child funerary jars help tell the story of Sub-Saharan Africa. Infant burial sites—including sacrificial altars—at "Old Chichen" (i.e. South America on the Mayan map) suggest South American infanticide as well. A drawing of the Heredia et al. jar is presented below.

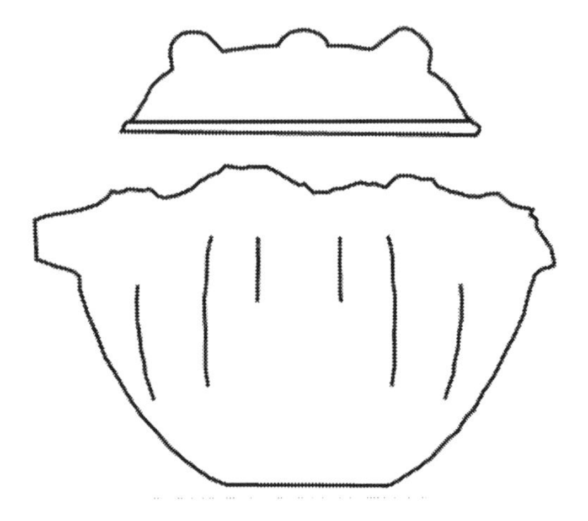

INFANT BURIAL CONTAINER FOUND DURING MAYALAND HOTEL CONSTRUCTION, CHICHEN ITZA. IMAGE ADAPTED FROM HEREDIA, CANUL, RUIZ, OSORIO AND ARIAS.

REMAINS OF A SUB-SAHARAN AFRICAN TEMPLE (REPLICA), MAYALAND HOTEL GROUNDS, CHICHEN ITZA.

TREE GROWING ON TOP OF THE REMAINS OF A SUB-SAHARAN AFRICAN TEMPLE REPLICA ON THE MAYALAND HOTEL GROUNDS AT CHICHEN ITZA.

The Middle East:
Israel and Sinai

Chapter 29

Israel and the Sinai Peninsula

STEAM BATH (ISRAEL REPLICA), CHICHEN ITZA.

On the Mayan map, Israel is referred to as the "Steam Bath." It is located roughly 50 meters—across open space—east of the Egyptian colonnade. The open space represents the Sinai Desert. According to the Bible, it is where Moses encountered the burning bush and received the Ten Commandments. Archaeologists describe the outer part of the Chichen Itza Steam Bath as a waiting room, the second part as the place where hot stones heated water, and the third part was where one bathed. Thus, the general description is consistent with the ancient Israeli temple in Jerusalem. First was the Outer Court, then the Holy Place, then the Holy of Holies—three sections. In other words, the Israeli temple replica includes a place where the priests could purify themselves by bathing, a place for burnt offerings, etc.

There are also unexcavated temples at Chichen Itza that are roughly located where Arish, in Sinai is located, and where Cyprus is located.

Note: A second Steam Bath on the Mayan map is found in the Southwestern United States near the Chaco Canyon replica. This is consistent with the theory that ancient Semitic people visited North America.

The Atlantic Region

Chapter 30

The Azores

Chichen Itza's "Grand Plaza" represents the Atlantic Ocean. It is west of the Temple of the Warriors ("France") and east of the Great Ball Court ("Washington, D.C."). The Atlantic Ocean replica appears most clearly after a heavy rain. Visitors fail to see it because they avoid the heavy rains, which turn the plaza into a replica of the Atlantic Ocean, and the Pyramid of Kukulkan into Mt. Pico in the Azores. As archaeologist Linda Schele suggested, the Mayan pyramids and plazas represent mountains and seas, respectively.

The Pyramid of Kukulkan is the largest and most elaborate structure at Chichen Itza to indicate the importance of the Azores in the Western Hemisphere. The reason for this importance is that, while the Prime Meridian currently runs through Greenwich, England, it previously ran through the Azores. The Azores had been chosen because they offered the most stable magnetic compass readings for travel. However, for political reasons—and contrary to France's preference—the Western nations voted to reestablish the Prime Meridian through Greenwich, Paris, etc.

The architect indicated the Azorean Prime Meridian by aligning it with Chichen Itza's largest sacbe, which runs north from the plaza to the Sacred Cenote (whose center represents the North Pole). Since—according to the map theory—sacbes represent magnetic fields, this largest of sacbes represents the strongest magnetic field in the Western Hemisphere—nature's Prime Meridian through the Azores. In other words, the architect used the Pyramid of Kukulkan to indicate the importance of the Prime Meridian in Western cartography.

In 2015, archaeologist Dr. Andres Tejero-Andrade discovered a cenote beneath the Pyramid of Kukulkan. Since it is covered with several meters of limestone, it is essentially an underground watery cave. Thus, it represents the watery caves ("grutas" or "grottos") beneath Mt. Pico. The most popular of which is "Gruta das Torres."

The 2016 discovery by Chavez and Tejero of another pyramid, in the volcano replica, represents the second volcano on the island of Pico—in the Madalena Volcanic Complex. Madeira (1998) and others have mentioned this second volcano.

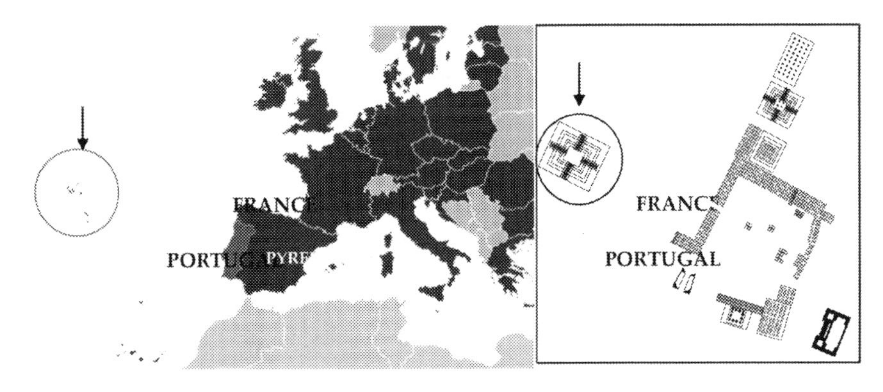

LEFT: MT. PICO AND AZORES (ENCIRCLED), PHOTO BY TYK. RIGHT: PYRAMID OF KUKULKAN (MT. PICO).

MT. PICO. IMAGE BY RAFAL GADOMSKI.

MT. PICO. IMAGE BY HENNER DAMKE.

PYRAMID OF KUKULKAN, CHICHEN ITZA. IMAGE BY AUTHOR.

Chapter 31

Azores: Stage One of Azorean Religion - Nature

The Pyramid of Kukulkan actually consists of three separate temples. According to the map theory, they represent three stages of religion in the Azores, which, presumably, represent the heart of the Western Hemisphere.

The first of the three temples—the hidden pyramid beneath the outer pyramid—represents the volcanic Mt. Pico. There is a lava-colored jaguar throne in it, as well as (previously) two human skeletons behind the throne. They presumably represent sacrifices to a volcano god (nature) in the ancient Azores. The Chac Mool who looks away from the lava-colored throne may represent the Azoreans turning away from the old religion to a new one—the worship of Tanit (discussed later on). The weakness of the volcano god hypothesis, however, is the lack of evidence of an Azorean society prior to Tanit worship.

Lava Channels

A Lava channel forms when lava flows from a volcano and burns a groove into the earth, along which more lava then flows. As the channel burns downward through rock and soil, any debris on top eventually collapses inward. As the hot lava continues on its path, it essentially bores a hole through the earth, leaving a stone tunnel in its wake. Accordingly, the images below show a complex of lava channel replicas adjacent to the Pyramid of Kukulkan, which represent the lava tunnels at Mt. Pico in the Azores. Note: most of the lava channel replica tributaries have been filled back in since 2009. However, they can be seen in one of the images below.

LEFT: LAVA CHANNEL REPLICA, CHICHEN ITZA.
IMAGE BY AUTHOR.

LEFT: LAVA CHANNELS. PHOTO BY KONSTIK | DREAMSTIME.

Chapter 32

Azores: Stage Two of Azorean Religion - Tanit

During the second stage of religion, the Azoreans worshipped Tanit from the Mediterranean region. The evidence is that in 2011, archaeologists Nuno Ribeiro and Anabela Joaquinito found ancient Azorean temples dedicated to Tanit, who was known as "Astarte" among the Greeks and as "Ishtar" ("Easter") in the Middle East.

Symbolized by twin snakes, Ishtar's mythological descent into the earth on the spring equinox, followed by her return, represented rebirth and resurrection. The Mayan architect presented this original Easter as twin snakes crawling down the Pyramid of Kukulkan—in light and shadow—on the spring equinox. The same phenomenon does not occur on the fall equinox because snakes do not typically crawl backwards uphill, and there are no snakes on the south side of the pyramid to capture the sun's southern trek.

Additional evidence that Kukulkan's pyramid represents the Azores is that in 1932, archaeologists found obsidian and coral objects in a small box in the pyramid. Coral represents the ocean (i.e. surrounding the Azores), and obsidian is found at volcanic sites. This supports the map theory, as the Azores are volcanic islands.

ISHTAR AND HER TWIN SERPENTS.
PHOTO MY BIBLELANDPICTURES | DREAMSTIME.

Chapter 33

Azores: Stage Three of Azorean Religion - Christianity

The third temple at the Pyramid of Kukulkan is located at its summit. This "Upper Temple" uses light and shadow (explained previously) to show Kukulkan as a god-man exiting a small room, at sunrise, on Easter. Kukulkan's ally (also in human form) is a warrior who guards the temple's back (southern) door. Evidence that the Upper Temple represents the garden tomb of Christ is that Ah Puch—the Mayan god of death (in the form of a skeleton)—is standing back in the darkness. In Mayan mythology, Ah Puch rules Mitnal, the worst part of the nine hells. His position in the darkness represents his mythic nocturnal hunt for victims. Thus, Christianity is the third stage of religion in the Azores—brought by the Portuguese. Roughly 80% of Azoreans are Roman Catholic. Accordingly, when gatherers at Chichen Itza's Grand Plaza clap their hands (to create an echo) before the Pyramid of Kukulkan on the spring equinox, they are inadvertently participating in a Christian Easter pageant. They either represent the visitors to the empty tomb, or modern congregants on Easter—often the most attended day of the year. Similarly, according to acoustic engineer David Lubman, the echo at the pyramid creates the chirping sound of the quetzal bird. The Aztec name, "Quetzalcoatal" (based on the bird) is known as "Kukulkan" among the Mayas. Thus, the chirping echo from clapping hands at the Pyramid of Kukulkan symbolizes the voice of Christ on Easter; hence, resurrection.

This does not mean the Mayas were Christians, but that the Azoreans would be—a thousand years after the pyramid was built. Since the Pyramid of Kukulkan is (apparently) the central structure of Chichen Itza, the Azores may represent the geographic and cultural focal point of the Western Hemisphere. If that is accurate, then the Christian Easter story (told at the crowning temple of Kukulkan's Pyramid) is the focal point of the Western Hemisphere. That is consistent with Christianity's status as the largest religion in the West.

The theory that Kukulkan (and Quetzalcoatl and Viracocha) represented Jesus (at least in some instances) is not new. It is part of a relatively old theory that Jesus visited the Americas. It is part of Mormon theology. While the theory presented here does not support or refute the idea of Christ in the Americas, Le Plongeon apparently believed Jesus visited the Yucatan. He said Jesus spoke "pure Maya" on the cross when he said, "Eli Eli lamah sabacthani?" This is typically translated as "My God, my God, why hast thou forsaken me?" However, Le Plongeon interpreted it as, "Hele Hele lamah zabac ta ni," meaning—"Now, now I am fainting; darkness covers my face."

There is clearly an overlap between Mayan and Christian symbolism, beyond what was mentioned previously in this discussion. For instance, one of the panels from Palenque shows a powerful man on the right, and a lower-status man (whose diminutive size requires that he stand on a box) on the left. Between them stands a bird standing on top of a large cross, which, in turn, stands on a skull. Each part is consistent with Christian symbolism (e.g. Christ's cross on top of "The Skull").

MAYAN CROSS ON A SKULL.

The Upper Temple matches St. Matthew's reference to a mountain from which Christ and the Devil could see "...all the kingdoms of the world at a moment in time." For instance, north of the Azores we see Iceland (the Platform of Venus) and the Inuit of the Artic Cordillera. To the south we find Balankanche, which represents Antarctica. To the east we see the European landscape replica, the Roman Empire, and Africa. To the west we see America and the sacred, Native American sites (e.g. Chaco Canyon, the Grand Canyon, etc.).

If the time traveler theory is accurate, then the architect may have seen Christ—in person. Thus, the carved image at Chichen Itza would be the world's only known image of Christ, based on first-hand observation.

LEFT: KUKULKAN (THE "GOD-MAN") AT THE NORTHERN DOORWAY, UPPER TEMPLE AT THE PYRAMID OF KUKULKAN. RIGHT: THE RESURRECTION OF CHRIST BY PETER RUBENS.

WARRIOR AT THE SOUTHERN DOORWAY AT THE UPPER TEMPLE, PYRAMID OF KUKULKAN, CHICHEN ITZA.

AH PUCH (MAYAN GOD OF DEATH) IN THE UPPER TEMPLE OF KUKULKAN, 1876.

Chapter 34

Iceland and Venus

The Grand Plaza's "Platform of Venus" is north of the Azores replica and represents Iceland. It is south of the Arctic Circle replica, and aligned with Washington, D.C. and Western Europe. This represents Iceland's close socio-political similarity to Europe and America, as opposed to Inuit society to the far north.

Iceland is called the Platform of Venus because, according to NASA, Iceland is the only place in the Western Hemisphere where one can observe both the beginning and the end of the Transit of Venus. This happens because Iceland uniquely falls in "Region X" –just above the sun's "X-path"—in the image below.

While the beginning of the transit can be seen with the naked eye at night, the end occurs after sunrise, and it can be seen with neither the naked eye nor with a telescope. Presumably, then, the Mayas learned about the transit from the architect.

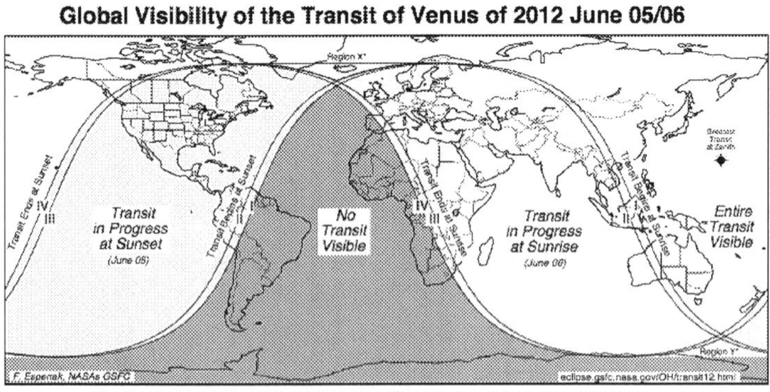

ICELAND IN "REGION X" TRANSIT OF VENUS.
IMAGE BY NASA.

Chapter 35

Iceland: Geography

The walls on the Iceland replica contain four-legged starfish and fish with fins. The starfish is similar to the symbol of Ishtar (see image below). The area in front of the platform once contained an engraved fish on a stone slab (see image below). These symbolize an ocean. The Iceland temple has three interior levels. The lower two are red—representing Iceland's red earth, and the upper one is white—representing snow and ice at Iceland's higher elevations.

STARFISH AT PLATFORM OF VENUS.

FISH IN BAS-RELIEF, PLATFORM OF VENUS,
IMAGE BY AUGUTUS LE PLONGEON, 1876.

Chapter 36

Iceland: The Island

In Meso-American culture an eagle and a snake on a rock means the rock is an island. We see the eagle killing the snake in the codices, in Aztec art, and on the Mexican flag. These symbols appear in a pageant of light and shadow at three temples at Chichen Itza: the two Platforms of Venus, and the Platform of Eagles and Jaguars. In the "platform" images below, the first one is not the Icelandic replica, but it has the same design. Note the eagle (in light and shadow) pursuing a winged snake with a long body in shadow. The image following it is the actual Platform of Venus (Iceland replica). At certain times—depending on the angle of the sun—that image also appears on the Iceland platform in the second image below. Thus, the architect combined Mayan mythology with light and shadow—as was done at Kukulkan's pyramid—to present a message.

PLATFORM OF EAGLES AND JAGUARS, PHOTO BY AUTHOR.

THE PLATFORM OF VENUS. IMAGE BY MILLAUS.

Chapter 37

Iceland: Keflavik Air Force Base

The "Cones"

The inside of the Platform of Venus once contained 182 "cones" as archaeologists described them. According to Alice Le Plongeon the cones ranged in length from roughly 1 foot to 4 feet. Since the average weight of a cone appears to be roughly 70 lbs., the collection of cones weighed about five tons. As that is quite a substantial amount, presumably the architect was making an important point.

According to Le Plongeon, two-thirds of the cones were painted blue, and one-third were painted red. It isn't clear where the larger cones are currently located, but some of the smaller ones are adjacent to the Platform of Venus at coordinates 20.684136, -.88568511. Archaeologists do not know the purpose of the cones, or why they have not been found in any other Mayan cities. They don't appear to represent tools or phalli, as they are not anatomically correct (as are the phalli at Uxmal).

The cones represent American military ordinance—bombs. The smaller ones are the size and shape of W-33 8-inch nuclear warheads. The faded red and blue colors are consistent with Iceland's—and America's—partially red and blue national colors. In other words, the architect is showing us a nuclear arsenal. Accordingly, the New York Times (Oct. 19th 1999), the Washington Post (Oct. 20th 1999), other sources indicate that the Freedom of Information Act revealed that the United States maintained a Cold War nuclear arsenal in Iceland. Note: Alice Le Plongeon indicated that the cones were found stacked in piles three and four deep—as bombs are currently stored.

"CONES" FROM THE PLATFORM OF VENUS. PHOTO BY AUTHOR (2014).

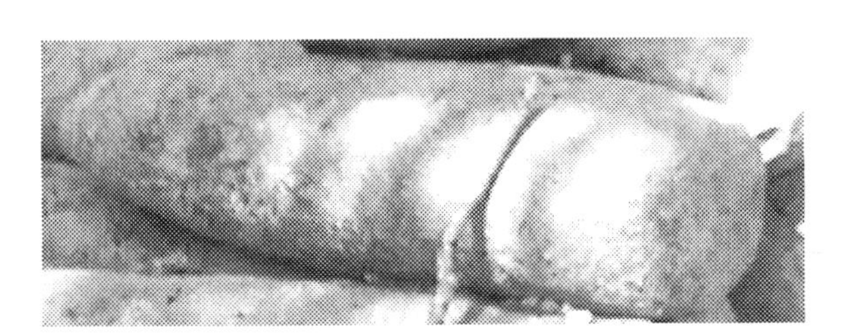

CONE FROM PLATFORM OF VENUS, CHICHEN ITZA.

The Tenoned Serpents

The Le Plongeons also found twelve tenoned, feathered, serpents in the Platform of Venus. Feathers suggest flight, and tenons mean the object is to be inserted into another object. Alice Le Plongeon—who photographed the initial excavation in 1873—said the tops of the serpents' heads were green, underneath they were yellow, and they had blue circles around their eyes. Their tongues were red and their eyes were white.

This is more symbolism. The inserted flying snakes represent WW II Air Force pilots at Iceland. The U.S. pilots wore green officers' hats with beige (similar to yellow) shirts. The blue circles around their eyes presumably mean the "flying snakes" see blue when they look around during flight—the sky.

The Mayas often used symbols. For instance, Kukulkan is a "bird-snake" because the former represents the heavens and the latter represents earth, as one can't get much closer to the earth than a snake. Also, as the image below suggests, the serpents (pilots) at the air base are not identical. For instance, some have their mouths open wider than others, and they have more "decoration" around their mouths. They presumably represent higher-ranking officers.

The serpents' snout antennas could indicate the ability to receive information through the air (i.e. radio waves), just as snakes use their tongues as antennae to sense prey, etc. Supporting evidence for the radio hypothesis is that no other serpents at Chichen Itza have antennas.

LE PLONGEON AND MAYA WORKERS WITH STONE SERPENT HEADS FROM THE TOMB OF CAY (PLATORM OF VENUS-NORTH).

Chapter 38

Iceland: Radiation

The Chac Mool Sphinx

The Iceland temple once supported a reclining Chac Mool Tiger on top of it, and which an early Mayan archaeologist referred to as a "wounded tiger." Since the tiger has a human head, it is technically a sphinx. However, unlike typical sphinxes, this one is wearing a sun hat. The evidence is its similarity to the hats being worn by the Mayan diggers in the image.

The Nordic nations—including Iceland—have naturally-high levels of radiation because of their proximity to the North Pole. Thus, Icelanders have a relatively high risk of exposure to harmful radiation from the sun. Hence, the only known sphinx to wear a sun hat.

CHAC MOOL AS A SPHINX/WOUNDED TIGER,
PLATFORM OF VENUS-NORTH, CHICHEN ITZA (1880'S).

The Standard Bearer

The Le Plongeons also found a statue of a boy inside the Platform of Venus. When the statue was unearthed, Alice noted that it was "thickly coated with loose mortar." She named him "Standard Bearer" because his right hand was in a clutching position (e.g. for a flagpole).

The boy's contorted posture suggests physical discomfort. His right foot is turned in, his head is turned left, and his left hand is in a defensive position. His lifelike eyes and facial expression suggest shock or fear. His detachable fingernails (made of sea shells or bone) have fallen off and have been lost. We only know of them because of Alice's field notes.

Since detached fingernails are a symptom of radiation sickness, perhaps the loose mortar on the boy represented nuclear ash. This might explain why he was displayed on top of the 182 warhead replicas. According to experts, this boy or perhaps young man was in the military, which is consistent with his location at an arsenal.

THE STANDARD BEARER – INTERIOR, PLATFORM OF VENUS-NORTH, 1880'S. PHOTO BY ALICE DIXON LE PLONGEON.

The Americas:
The United States

Chapter 39

Washington, D.C.: The National Mall

There are more pillars on the European side of the Mayan map than there are on the American side, because Greco-Roman architectural influence was stronger there. Thus, while pillars are the most outstanding feature at the replica of Europe, the most outstanding feature on the American side is a replica of the National Mall in Washington, D.C. The miniature D.C. replica includes the Quadrangle, the White House, the Capitol Building, the Lincoln Memorial, the Vietnam Wall, the Smithsonian Museum of Natural History, the capital area fallout shelter, and Kalorama.

Since Chichen Itza wasn't excavated until a century after Washington, D.C. was laid out (in the late 1700's), it isn't possible for Washington to be a copy of Chichen Itza. And, aside from time travel, it is difficult to explain how the Mayas could have copied Washington, D.C., since copies can't precede the originals. Thus, while it is not clear what Fred Hoyle's evidence of time travelers was, his conclusion is consistent with the data at Chichen Itza. The architect appears to have had no time or space restrictions.

THE NATIONAL MALL, WASHINGTON, D.C.
TOP: CAPITOL BUILDING, CENTER: WASHINGTON MONUMENT, CENTER-LEFT: SMITHSONIAN MUSEUM, BOTTOM: LINCOLN MEMORIAL.

Chapter 40

Washington, D.C.: The U.S. Capitol Building

The U.S. Capitol Building replica remains unnamed. However, it shares three features with the Capitol Building. First, its interior arrangement of pillars is similar to that in National Statuary Hall in the Capitol Dome. The pillars at the capitol building have statues of men standing before them, while the pillars at the Chichen Itza structure have images of men carved directly on them. Second, this structure is located on the opposite end of the quadrangle from the Lincoln Memorial replica. Third, the contorted body (simultaneously sitting, laying, turned and straight) of "Swimmer Chac Mool," who once stood in the Capital Building replica, presumably represented political compromise.

Solar Calendars

In 2012, archaeologists discovered that the structures on top of Chichen Itza's Great Ball Court indicate the equinoxes and the solstices. In other words, it was a solar calendar. This is consistent with the map theory, as Washington, D.C.'s Pennsylvania Ave., Maryland Ave., and Capitol St., provide the same function: At sunrise on the summer solstice, the sun shines down Maryland Ave. and illuminates the Capitol Building; on the winter solstice it shines down Pennsylvania Ave. and illuminates the Capitol, and on the equinoxes it illuminates the Capitol from Capitol Street.

UNNAMED TEMPLE AT THE GREAT BALL COURT.

NATIONAL STATUARY HALL, U.S. CAPITOL BUILDING.

Chapter 41

Washington, D.C.: The National Mall Quad

Mayan ball courts all look somewhat different because each court represents a different fault line. Chichen Itza's "Great Ball Court" represents the quadrangle at the National Mall. It is near the Quail Fault Line in Virginia, which caused the earthquake that damaged the Washington Monument in 2011. The quadrangle is also known as "The Green."

As explained previously, the stone rings at Mayan ball courts represent planetary, cross-sectional, earthquake models—maps. The rings at Chichen Itza's Great Ball Court use intertwining snakes to show primary and secondary shock waves. The thin outer ring represents the earth's crust, the hollow interior represents earth's inner core, the striking serpents show the hypocenter (beneath the epicenter), etc.

THE GREAT BALL COURT RING, CHICHEN ITZA. PHOTO BY KONSTANTIN KALISKO.

Voice Amplification

Archaeologist Chris Scarre said Mayan plazas were often "stage sets" in a theatrical sense. This is consistent with the pageant hypothesis. Theatrics require voice amplification, which—according to National Geographic (2010)—the Mayas knew how to do. Thus, the Great Ball Court (replica of the quadrangle) was designed to amplify speech (via its high walls) to help tell the story of Washington, D.C. The quad is where presidents are sworn into office, and where Dr. Martin Luther King gave his "I have a dream" speech, etc. They all involve amplified speech (e.g. microphones, loudspeakers, etc.).

AMPLIFIED SPEECH AT THE U.S. NATIONAL MALL-MARCH ON WASHINGTON. PHOTO BY WARREN K. LEFFLER.

AMPLIFIED SPEECH AT THE U.S. NATIONAL MALL-PRESIDENTIAL INNAUGURATION, PHOTO BY THOMAS MENEGUIN.

AMPLIFIED SPEECH AT THE U.S. NATIONAL MALL-BRITNEY SPEARS.

Chapter 42

Washington, D.C.: The Lincoln Memorial

The structure opposing the Capitol Building replica on the Mayan map represents the Lincoln Memorial. It is called the "Temple of the Bearded Man" because that is how the Mayas (who didn't have facial hair) described men with facial hair. For instance, in more modern times they referred to Augustus Le Plongeon—not by his given name—but as "Long Beard." Clearly (see image farther below), the bearded man at the replica is European rather than Mayan. Moreover, the two plumes over his head mean he was in a war, and his visible, final, "breath of life," means he died prematurely. These cues are consistent with the story of Abraham Lincoln. The hieroglyphics at the temple are about a battle, just as the Lincoln Memorial displays the Gettysburg Address from the Battle of Gettysburg. Moreover, the Lincoln Memorial and the National Mall are known for voice amplification (see images below), just as the Great Ball Court was designed for the same purpose.

TEMPLE OF THE BEARDED MAN,
GREAT BALL COURT, CHICHEN ITZA.

LINCOLN MEMORIAL, U.S. NATIONAL MALL,
WASHINGTON, D.C.

TEMPLE OF THE BEARDED MAN, CHICHEN ITZA.
PHOTO BY AUTHOR (2013).

LINCOLN MEMORIAL, NATIONAL MALL, WASHINGTON, D.C.,
(TEMPLE OF THE BEARDED MAN).
PHOTO BY CHRISTOPHER ENG-WONG.

MAYAN GLYPHS ABOUT A BATTLE, SHOWN AT THE TEMPLE OF THE BEARDED MAN.

LEFT: ABRAHAM LINCOLN. RIGHT: THE BEARDED MAN TAKING HIS LAST BREATH, TEMPLE OF THE BEARDED MAN, CHICHEN ITZA. PHOTO FROM 1800'S.

It seems unlikely that a random collection of stone monuments would accurately present so many disparate people, places, and events. Rather, it appears that the architect knew more than any scholar or clairvoyant that we know of. To apply a concept raised previously, did the architect simply know—a thousand years in advance—that Lincoln would be memorialized? Did he know Lincoln was going to be elected president? Free the slaves? Assassinated? Was the architect the type of being about which Fred Hoyle warned us… a time traveler whose abilities greatly exceed our own, and who influences our decisions? Or, were Hoyle's critics right, and the Big Bang theorist (and Nobel Prize candidate) was a genius who lost touch with reality? If not, then this all raises questions of freewill, fate, determinism, mind control, etc., as discussed previously.

Chapter 43

The Lincoln Memorial: Coin Sculpture

Beneath the face of the Bearded Man lies a stone disk object about two feet in diameter, and roughly eight inches tall. A side hole, inches from the top, guarantees drainage and a raised perimeter. Since the surrounding temple was designed without a complete roof, the apparent effect is as follows: On certain days and nights, rainwater that has collected in the sculpture reflects a disk of sunlight or moonlight onto the face of the Bearded Man. The effect would resemble an American penny. That is why Lincoln is shown in profile at the temple.

LEFT: DISK AT THE TEMPLE OF THE BEARDED MAN.
RIGHT: AMERICAN PENNY (ABRAHAM LINCOLN).

Chapter 44

Washington, D.C.: The Vietnam Wall

Exiting the Lincoln Memorial at an angle to the left places one at the Vietnam War Memorial Wall. It is inscribed with the names of fallen soldiers. The Mayan map has the same arrangement. As one exits the Temple of the Bearded Man at an angle to the left, one encounters the "Tzompantli," a stone wall of fallen Mayan soldiers. It is T-shaped and displays soldiers' skulls rather than their names. Presumably the Mesoamerican T-shape represents death, as it is carved on the body of the death goddess.

Chance, alone, does not adequately explain these similarities. As stated previously, the map raises questions about freewill and fate. The architect was either simply aware of the future, or he caused it—the war, the memorial's design, its location, etc.

VIETNAM WALL OF DECEASED SOLDIERS,
WASHINGTON, D.C., IMAGE BY AUTHOR.

WALL OF DEAD SOLDIERS ("TZOMPANTLI"), CHICHEN ITZA.
PHOTO BY BJORN CHRISTIAN TORRISSEN.

MESOAMERICAN GODDESS OF DEATH, NATIONAL
MUSEUM OF ANTHROPOLOGY, MEXICO CITY.

Chapter 45

Washington, D.C.: The White House

The White House replica on the Mayan map is called the Platform of Eagles and Jaguars. The name derives from a set of engraved panels—an eagle next to a jaguar—on the structure's south side. Each animal is shown eating a human heart, which presumably indicates great social-political power, particularly since each animal is at the top of its respective food chain. Maya rulers were called "Jaguar Priests," while the symbol of American rulers is an eagle—the presidential seal. Presumably, the squirrel eating the human heart at the "France" temple does not mean squirrels are at the top of the Parisian food chain. Rather, it probably refers to a comparison of U.S. versus French influence.

PLATFORM OF EAGLES AND JAGUARS. PHOTO BY AUTHOR.

LEFT: EAGLE, PLATFORM OF EAGLES AND JAGUARS, CHICHEN ITZA (1800'S). RIGHT: SEAL OF THE PRESIDENT OF THE UNITED STATES.

JAGUAR, PLATFORM OF EAGLES AND JAGUARS, CHICHEN ITZA (1800'S).

AUGUSTUS LE PLONGEON AND MAYAN WORKERS WITH CHAC MOOL, FROM THE PLATFORM OF EAGLES AND JAGUARS (1800'S).

The White House Secret Service

The architect gave the White House the same design as Iceland to indicate that is an island. However, unlike the Icelandic replica, which is surrounded by fish, the White House replica is surrounded by men with modern weapons in sniper positions. There are sixteen of them, presumably enough for a first line of defense. A former U.S. Army soldier described the (full version of the) image below as "*two soldiers in combat helmets on guard duty wearing night vision goggles and holding M-1 rifles. They are on the lookout in a defensive posture… Also, each is defending the other's back.*"

Some of the soldiers in the images appear to be holding rifles, while others seem to be holding rocket launchers. If the former soldier's interpretation (above) is basically correct, then the architect knew there would be Secret Service men on top of the White House a thousand years before the White House was built, before America existed, etc. He knew how many snipers there would be on the roof, what kind of gear they would have, and where they would be concealed, as evidenced by cloud volutes over the images at the replica. It is unlikely that random chance duplicated reality here.

Sharpshooter Details

The first image, below, shows a tiny soldier (wearing a backpack), ostensibly hanging from the barrel of a secret service agent's weapon. The tiny soldier's size suggests he is far away in the sky. The tip of the agent's weapon on the right side of the image (held by an unseen soldier) is crossed with the sickle-like weapon of Ah Puch, the Mayan god of death. Thus, the architect is indicating that the object is deadly. In the second image below we see a different agent. His weapon is slightly different from the first soldier's weapon. The image of the third soldier is upside down. Since this photo appears in Le Plongeon's book from the late 1800's, it is evidence that the carvings are not modern fabrications (e.g. the work of vandals). The fourth image below show the right side up version of the third image. Finally, the fifth image is of an actual Secret Service agent—with a weapon—concealed on the roof of the White House.

ROOFTOP COUNTER-SNIPER, PLATFORM OF EAGLES AND JAGUARS, IMAGE BY SANTIAGO CAIZAPANTA.

ROOFTOP COUNTER-SNIPER, PLATFORM OF EAGLES AND JAGUARS. IMAGE BY MS DEBORAH WATERS.

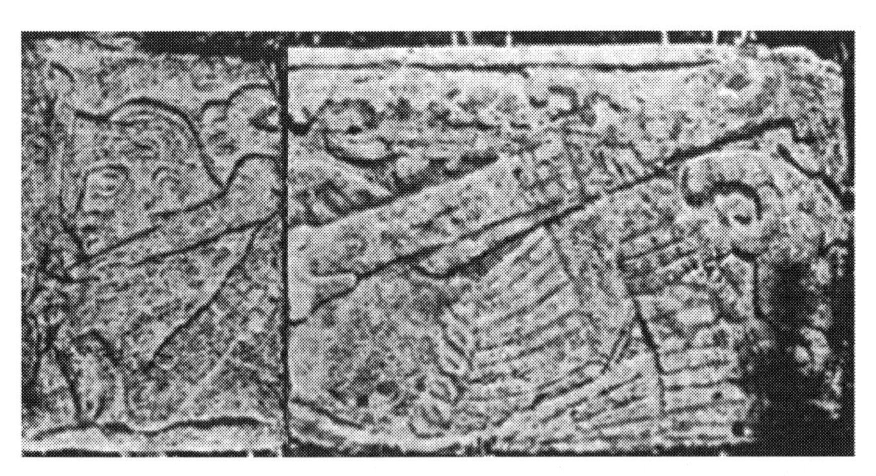

ROOFTOP COUNTER-SNIPERS, PLATFORM OF EAGLES AND JAGUARS. 1896, BY ALICE AND AUGUSTUS LE PLONGEON.

ROOFTOP COUNTER-SNIPERS, PLATFORM OF EAGLES AND JAGUARS. PHOTO BY AUTHOR (2016).

ORIGINAL GOVERNMENT TITLE: "SECRET SERVICE ON WHITE HOUSE ROOF." SOURCE: UNITED STATES DEPARTMENT OF HOMELAND SECURITY.

Chapter 46

Washington, D.C.: The Smithsonian Museum

The Mayan map includes a structure where the IRS headquarters and the Smithsonian Museum of Natural History are located. The structure is called the "House of the Jaguars." This suggests a government institution, as the Mayan jaguar priests were the leaders of society.

Alice and Augustus Le Plongeon copied several images from the Jaguar House, which are presented below. Generally, Augustus interpreted the images as Queen Moo's nuptials, and archaeologists suspect Queen Moo was mythical. In light of the map theory, however, the remaining alternatives are the IRS and the Smithsonian Museum of Natural History. Alice's images suggest the answer, as they show groups of people studying dead animals and pottery under the guidance of what appear to be instructors.

Augustus referred to the Smithsonian replica as "Prince Coh's Memorial" chamber. He described its paintings and carvings as geographically diverse, representing the Americas, Egypt, Africa, Angkor, Burma, etc. Scenes below show teachers (i.e. scholars, tour guides, etc.) pointing at an armadillo, a pig, a fish, a piece of pottery, etc. against a backdrop of huts, which may represent display booths. Le Plongeon suggested that, while Coh's temple contained no hieroglyphics, it explained much with images. He said the temple largely told the story of a little girl named "Moo," as she progressed from one stage of life to the next. She apparently became the elderly instructor in one of the scenes.

Le Plongeon's evidence for Moo's existence is weakened, in that it is partly metaphysical. He believed he and his wife, Alice, were the reincarnated Prince Coh and Queen Moo, although the prince and queen had been siblings. Thus, in light of the map theory, an alternative hypothesis is offered. Queen Moo may have metaphorically referred to earth, and her presence in the Smithsonian replica images may have referred to mankind's progress through the ages. This hypothesis is partially supported by a particular painting in the Smithsonian replica, which seems to show three stages of man.

The Journey of Man

In the first stage we see a man emerging from the waters of a marsh. He is wearing little or no clothing. He seems to be carrying a fishing net and a long bundle of sticks. He is at very low elevation as he climbs up a steep embankment past an apparent Procoptodon at the shore. According to Smithsonian magazine, Procoptodons lived in South America before they migrated to Australia, where their faces grew longer, and they became modern day kangaroos.

In the Mayan image, the Procoptodon sees the venomous snake coiled around the tree behind him. However, unlike the kangaroo, the man in the image has avoided that particular snake, as well as another one who is attacking an unsuspecting bird on a tree top. A koala hugs the tree trunk beneath the bird, and beneath the koala rests a large snake. It is coiled four or five times around, and appears to be eyeing the nearby village. Since, according to experts, koalas hug trees to keep cool, this environment was presumably hot. Moreover, since anthropologists indicate that both kangaroos and koalas originated in South America, it is plausible that the Mayas of Central/North America knew of them. It is unclear—from Alice Le Plongeon's reproduction of the thousand year old mural, however—whether the kangaroo has a short face and long ears, or a long face and short ears.

In the second stage of the image, the man has emerged from the forest, having left behind its orange and apple trees, and its dangers. He has reached a (presumably metaphoric) plateau, and is now fully-clothed in a toga-like garment. Thus, he is transitioning from living in a forest/jungle among the other animals, to living in a house. There are no trees or animals on this side of the image. Thus, in the third stage, man has left his Hobbesian jungle and has become more social. There is an abundance of food, evidenced by the bowl of "apples" and "oranges" according to Alice Le Plongeon.

According to experts, prehistoric koalas and kangaroos arrived in Australia roughly 80 million years ago, when South America, Africa, Antarctica and Australia formed the larger part of a supercontinent known as "Gondwana." Today there are no indigenous kangaroos and koalas in South America, and it is apparently difficult for scientists to determine when they became extinct there. However, if we assume the South American kangaroos and koalas all migrated to Australia at the same time, then they became extinct in South America 60 to 80 million years ago. This, of course, would mean humans lived alongside koalas and kangaroos 80 million years ago. The problem with this hypothesis is that, according to scientists, mankind appeared on earth no longer than 6 million years ago. If, on the other hand, the Mayan image refers to Australia, then it means Pre-Columbian Mayas somehow knew about Australia. If the former is accurate, then it supports the time travel hypothesis. And, if the latter is true (and the Mayan image describes Australia) then it supports the map theory.

Other images at this particular temple show people among a series of huts (booths), each performing a daily task (carrying objects, sitting and talking, etc.), a Native American man in a canoe, and what appears to be a Greco-Roman soldier. Thus, the contents of Prince Coh's Memorial are consistent with the subjects and themes actually found in the Smithsonian Museum of Natural History.

The Mayan map also accurately shows the Natural History Museum's two entrances; one on the White House side (with a jaguar throne), and the other on the quad side between the Capitol Building and the Lincoln Memorial replicas. On the Mayan map, the museum's quad-side entrance is elevated and aligned with the Ball Court wall. Presumably this facilitates the acoustics for amplified speech (discussed earlier). In other words, Prince Coh's Memorial represents the mall-side entrance to the Smithsonian Museum, and the House of the Jaguars represents the museum's entrance that faces the White House. Overall, the meta-analysis facilitated by including a museum of earth's history is consistent with a hall of earth's records. In other words, the architect placed a miniature structure with historical images of nature in it, in the same place where the Smithsonian Museum of Natural History is located.

There is more evidence. The trio of images, below, shows three European-looking men. The man on the far left has a lion over his head and on his chest. Thus, he represents a high-ranking leader. The second man, also European, may represent a politician or social scientist; he seems to literally have people on his mind. The third man is a stern-looking warrior with a grail (cup) over his head. This may represent the Holy Grail or some other cup of historical significance. A brief argument (largely speculation) in favor of the Holy Grail hypothesis is as follows.

The Holy Grail

At least one version of the grail legend ends in America. It purports that the grail came to America by a courier-priest, on the same ship as the legendary army Captain John Smith. The legend does not say Smith was protecting the grail, or that he even knew it was aboard ship. However, it seems strange that two such outstanding figures would be sailing on the same ship, especially given that Smith had previously worked for the Hapsburgs, who were part of the Merovingian lineage and the story of the grail. It is also curious that Smith committed some sort of crime aboard ship—the night before they reached shore. One minute he was to be hanged, and then next minute—after a sealed container was opened—he was declared Colonial Governor of Virginia. Meanwhile, the priest—according to the legend—disappeared among the Accokeek Indians of the modern-day Washington, D.C. area. Accokeek is across the Potomac River from Popes Creek Plantation, where George Washington would grow up years later. As a surveyor, he would have been familiar with the local landscape, and possibly with its legends. There is no evidence that George Washington had the Holy Grail. However, historians have noted his "luck" in battle, having survived a number of near-misses. It is also interesting that the nation's capital was located in a Y-shaped (grail-shaped) geographic region (because of the local rivers), having been carved from Virginia and Maryland. Symbolically, this makes the city "Virgin Mary land," which, admittedly with some stretch, is a grail reference. Perhaps that is all coincidental. However, in light of that suspicion, it raises some questions about the secret society George Washington co-founded at the end of the Revolutionary War. It also raises some questions about a member of Washington's secret society (or, a society with secrets) named Arthur St. Clair, and his role in the grail story. St. Clair had been President of the United States prior to George Washington (and, later a member of George Washington's secret society), having arrived in the New World from the Merovingian St. Clairs of Scotland, a remnant of the Templar massacre in France. The issue is that St. Clair massacred many Native Americans, and he may be the reason why the Accokeek Indians no longer exist. All that remains of them is the modern-day town of Accokeek, Maryland. In other words, St. Clair may have been searching for the grail.

It is also curious that the artist of the Mayan image chose to place the grail over the man's head, instead of, for instance, at his chest. While it is true that lions can't fly (i.e. the first image), the grail legend and its depictions tend to elevate it. Several masterpieces show the grail aloft, including the "Knight of the Holy Grail" at the Smithsonian. Da Vinci used a similar technique. In the Last Supper, he placed the grail cup over the head of St. Bartholomew, who sits on the far left side of the image. The identities of each person at the table appear in Da Vinci's original sketch of the scene. Note, however, that most reproductions of the Last Supper omit the grail. However, to continue, as Grandmaster of the Knights Templar, Da Vinci may have been indicating the cup's location at St. Bartholomew's Church in Haslemere, England. It is known for its stained glass images of the grail, and for the theft, in 1555, of a valuable chalice. In other words, decades later, it could have wound up on the ship with John Smith. This is all speculation. However, if the map theory is accurate, then the Mayan temple indicates that a cup of historical significance will be at the Smithsonian Museum at some point.

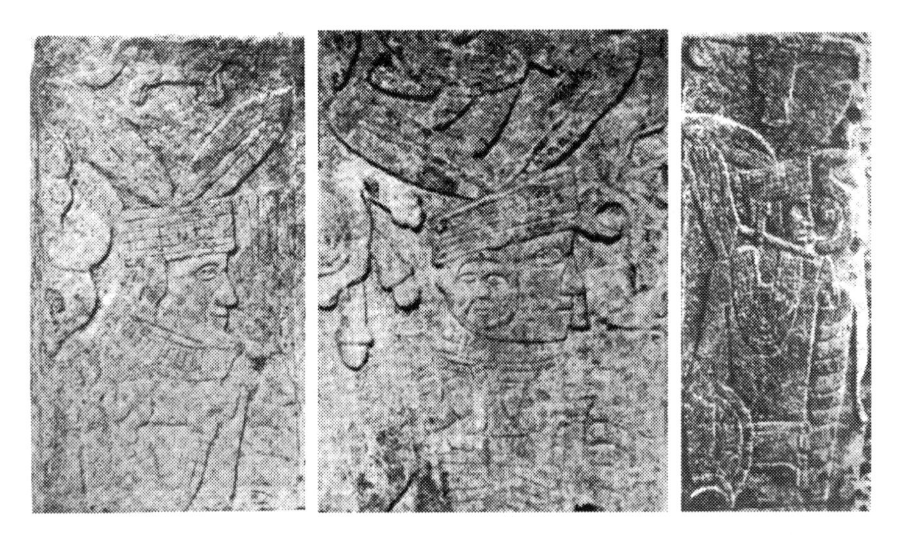

PILLAR CARVINGS AT THE UPPER TEMPLE,
HOUSE OF THE JAGUARS.

HOUSE OF THE JAGUARS. IMAGE BY ALICE LE PLONGEON.

IMAGE FROM UPPER TEMPLE OF THE HOUSE OF THE JAGUARS: STUDYING AN *ARMADILLO*, 1876.

IMAGE FROM UPPER TEMPLE OF THE HOUSE OF THE JAGUARS: STUDYING A FISH, 1876.

IMAGE FROM UPPER TEMPLE OF THE HOUSE OF THE JAGUARS: STUDYING A *PIG*, 1876.

IMAGE FROM UPPER TEMPLE OF THE HOUSE OF THE JAGUARS, 1876.

IMAGE FROM UPPER TEMPLE OF THE HOUSE OF THE JAGUARS: PEOPLE AT HUTS (DISPLAYS), 1876.

IMAGE FROM UPPER TEMPLE OF THE HOUSE OF THE JAGUARS: PEOPLE AT HUTS (DISPLAYS), 1876.

LEFT: KOALA IMAGE, COPIED FROM THE HOUSE OF THE JAGUARS, 1876.RIGHT: ACTUAL KOALA IMAGE, © PERCEPTION | DREAMTIME.

PREHISTORIC KANGAROO, HOUSE OF THE JAGUARS.

MODERN DAY KANGAROO.

MURAL, HOUSE OF THE JAGUARS.

SCENES FROM EVERYDAY LIFE, HOUSE OF THE JAGUARS.

SMITHSONIAN MUSEUM OF NATURAL HISTORY.
PHOTO BY SOOKSAN KASIANSIN.

WARRIOR IN A BOAT. IMAGE FROM TEMPLE
OF THE JAGUARS. BY LE PLONGEON.

GRECO-ROMAN SOLDIER REPLICA. IMAGE FROM
UPPER TEMPLE OF THE JAGUARS. BY LE PLONGEON.

LEFT: HOUSE OF THE JAGUARS
(BALL COURT ENTRANCE) BY LE PLONGEON.

Chapter 47

Washington, D.C.: Kalorama

The Washington, D.C. neighborhood of Kalorama (meaning "good view") is where the wealthiest residents of the capital district live. Kalorama is northwest of the National Mall. Similarly, there is a group of houses roughly 100 meters northwest of the Great Ball court, which Cyark archaeologists suggest was where Chichen Itza's elite lived. To repeat, the architect knew details about Washington, D.C. long before it existed. And, since the city was laid out a century before Chichen Itza was excavated, it is impossible that L'Enfant and Banneker—the men who designed Washington, D.C.—were inspired by it.

Chapter 48

Washington, D.C.: The American Revolutionary War

Myths Without Moral Lessons

Mayan myths are allegorical and geographically anchored in the map. Thus, unlike Aesop's fables, for instance, they don't convey a lesson or a moral. For example, Aesop's *The Fox and the Grapes* teaches us that flawed egos spurn what they fail to attain. However, such a story in the Mayan context would refer to a specific person. For instance, as will be discussed later, there is nothing kind or noble about 400 men plotting to murder a man who had just finished helping them—unless the story is allegorical. For, indeed, in light of the map theory, those 400 men were among the most important and courageous men in American history.

An American Emphasis

The next few sections suggest additional ties between the United States of America and the Mayan map. Consider the following. The highest level of Mayan heaven was called "Omeyocan," meaning "House of the Lord of Abundance." "Omeyocan" sounds like "American" being pronounced by a non-English speaker (e.g. an ancient Mayan). Similarly, a house of abundance is synonymous with America's moniker, "Land of Plenty." Finally, an American physicist (Dr. Ronald Mallett) is currently designing a rudimentary time machine. Thus, overall, the architect seems to be an American time traveler from our distant future.

The Great Ball Court and Washington, D.C.

The map theory presents Chichen Itza's Great Ball Court area as a replica of Washington, D.C. Therefore, any Mayan myths or rituals anchored in the Great Ball Court either refer to America or to Washington, D.C. Accordingly, America's colonial and independence eras are told allegorically in the Mayas' Popol Vuh, through myths connected specifically with the Great Ball Court.

Xibalba and England

The Mayan account of America's colonial era begins with the story of the "Maize God Twins." The twins loved playing on the Great Ball Court at Chichen Itza. However, they routinely made too much noise, which angered the evil lords who lived below ground in Xibalba, the "Place of Fear." The penalty for offending the Lords of Xibalba was decapitation, the burning of one's bones, and the pulverization of one's ashes. This punishment was very similar to the penalty for committing treason against the King of England in Colonial America. The Maize God Twins will be discussed in more detail below.

Three Rivers to "Xibalba"

According to the Popol Vuh, one reached Xibalba by navigating three rivers. The first was full of scorpions, the second was full of blood, and the third river was full of pus. This is more allegory. To reach Parliament from Washington, D.C., one first sailed down the Potomac River among the "fresh-water scorpions." Then, one navigated the Atlantic Ocean, which is red at sunrise—"blood." Finally, one sailed up the Thames River to the "Evil Lords of Xibalba," at British Parliament in London. The Thames was once so full of waste (i.e. "pus") that it spawned the modern public health movement. Students of the history of epidemiology begin at the Thames.

"THE SILENT HIGHWAYMAN" ON THE POLLUTED THAMES RIVER, 1858.

PALACE OF WESTMINSTER (BRITISH PARLIAMENT) ON THE THAMES BY BRYAN BUSOVICKI | DREAMSTIME.

The Maize God Twins

The Maize God twins represent Colonial America's leaders prior to the Revolutionary War. Many of those early leaders were Freemasons, for whom maize symbolized the dedication and constitution of a new lodge—America. Freemasonry was the source of many of America's ideals (e.g. free inquiry) and symbols.

Just as the Maize God twins did not succeed against Xibalba, however, so did the colonialists (prior to George Washington) fail against England. Similarly, the lifelike statues that the Lords of Xibalba used to trick the Maize God Twins represent the lifelike images of the Magna Carta Barons at the House of Lords.

Furthermore, the hot bench from which unsuspecting visitors suddenly leap, is allegorical as well. This mythic bench is located in the "Throne Room" of the Lords of Xibalba because it represents the literal benches in the British House of Lords, where the spectacle of leaping up from a bench is still played out. England's politicians spring up from their benches—as if from a hot seat—because only one may speak at a time, and whoever stands first has the floor. The throne in the Mayan myth refers to the monarch's (i.e. the Queen of England) chair, which is adjacent to the benches in the House of Lords.

Columbia and the Hero Twins

Although the Maize God Twins were executed, one of them subsequently impregnated the daughter of a Xibalban lord. He managed this by spitting onto her hand from his decapitated head (which was in a tree). This is clear evidence that the entire account is allegorical. However, to continue, the embarrassed young woman fled into isolation, where she gave birth to the "Hero Twins." Allegorically, she represents Columbia, the goddess of the New World who rebelled against the Old World in Europe. Her parents are the evil Lords of Xibalba—England. Her union with America's pre-revolutionary fathers (one of the Maize God Twins) produced the Revolutionary War era's founding fathers—George Washington and John Adams (the Hero Twins). Just as the pregnant woman's parents ostracized her in the Mayan myth, so did America become estranged from England. Today, a large statue of Columbia adorns the top of the Capitol Dome in Washington, D.C.

LEFT: GEORGE WASHINGTON CROSSING THE DELEWARE BY EMANUEL LEUTZE, 1851.
RIGHT: JOHN ADAMS (HAND ON HIP).

BRITISH PARLIAMENT, HOUSE OF COMMONS, 1793. PAINTING BY KARL ANTON HICKEL.

Vucub Caquix: King George III

According to the Popol Vuh, between the time of the Maize God Twins and the Hero Twins, there lived an evil bird-demon named Vucub Caquix ("Voo cub ca' kwish"). He sat on a thrown, wore jewelry, had teeth, and declared himself to be the sun and the moon; the sole ruler of the sky—a monarch. Vucub Caquix was not an actual bird, because birds don't have teeth, thrones, etc. Allegorically, he represents England's (American Revolutionary war era) King George III, whose porphyria symptoms (i.e. foaming at the mouth, etc.) were attributed to demon possession. Such was the common diagnosis of the day. However, the story of Vucub Caquix and the sets of twins is better understood in light of the sarcophagus of Lord Pakal at Palenque.

KING GEORGE III, BY ALLAN RAMSAY, 1762.

Twins versus Demons

Lord Pakal is referred to on an ancient stele from Chichen Itza, as well as on a sarcophagus at Palenque, which is actually a three-part puzzle. Its message is revealed when its borders are removed and the three parts are assembled. Pakal is thereby transformed into the two sets of twins from the Popol Vuh. Moreover, the Vision Serpent (from the World Tree) is transformed into the jaguar-crowned bird-demons Vucub Caquix and one of his son's, presumably Zipacna.

In the image below, Vucub Caquix is the bird-demon with old eyes in the tree on the left. His wings are curved as though he is flexing his biceps. This reflects his strength and prideful arrogance—having declared himself to be the sun and the moon. Beneath him, on either side of the tree, are the Maize God Twins. In the tree on the right, Vucub Caquix's son, Zipacna (also wearing a jaguar crown), has young, confident eyes. His wings are also flexed like arms, and his talons are splayed for attack as he flies downward between the Hero Twins.

Allegorically, Vucub Caquix is wearing a jaguar crown because he represents England's King George III. His "son" is wearing a jaguar crown because he symbolizes Lord Cornwallis—the king's general. The son (Zipacna) is in attack posture to symbolize England attacking America.

LEFT AND CENTER: MAIZE GOD TWINS WITH CROWNED BIRD-DEMON VUCUB CAQUIX (SEE ARROW). CENTER AND RIGHT: HERO TWINS WITH ZIPACNA (SON OF VUCUB CAQUIX) WEARING A JAGUAR CROWN – SARCOPHAGUS OF PACAL. SINGULAR IMAGE BY NICKU.

Houses of Xibalba

Ancient Mayan pottery shows the hero twins opposing the Lords of Xibalba. One side wears red hats while the other wears either blue or black. These represent the opposing British and American uniforms during the Revolutionary War.

In the Mayan myth, the evil lords of Xibalba invite the Hero Twins for a visit. Their intent is to trick and kill them, as had been done to their predecessors, the Maize God twins. To accomplish this, the evil lords presented the Hero Twins with a series of "tests" (obstacles) in "houses" to overcome.

Allegorically, the tests in the houses represent American Revolutionary War battles, as follows: "Fire House" represents the "Burning of Norfolk." "Cold House" represents winter at Valley Forge. "Razor House" (wherein many knives and swords are used), represents the Battle of Bunker Hill (i.e. the "Bunker Hill Sword"). "Bat House" represents a battle featuring "Bar Shot" shrapnel. Bar shot presumably flew from cannons with bat-like movements, cutting both ships' masts, and men, in two. That would explain Hunapu's—one of the Hero Twins—decapitation by a bat in the myth. Thus, the "blow gun" in the myth refers to a canon. Finally, "Dark House" may represent the inexplicable fog that suddenly concealed George Washington's retreat to Manhattan during the Battle of Brooklyn.

LEFT: WASHINGTON AND LAFAYETTE AT VALLEY FORGE, BY JOHN WARD DUNSMORE.
RIGHT: THE BURNING OF NORFOLK.

Chapter 49

The American Revolutionary War: The 400 Drunken Men

The reader is advised to read about the "Four hundred Drunken Young Men" in the Popol Vuh (free online) before reading this next section. Otherwise, it may be difficult to follow. The Popol Vuh describes 400 drunken young men who have a violent encounter with Zipacna—one of Vucub Caquix's (the bird-demon's) two sons. The myth prospectively describes America's first battle (the Battle of Brooklyn), in its first war—the Revolutionary War.

The Battle of Brooklyn (also called the "Battle of Long Island") began when British General Cornwallis (i.e. the "son" of Vucub Caquix) and his men crossed New York Harbor to the beach at Gravesend Bay and Gowanus Creek (river) in Brooklyn. America would have lost the ensuing battle and the war, save for the heroism of the Maryland Volunteers, known as the "Maryland 400."

In the Mayan myth, the bird-demon's sons represent the top two British military leaders in Colonial America—Lord Cornwallis and his second in command. That is because the bird demon represents King George III (explained previously). Zipacna's beach in the Mayan account represents Gowanus Creek, where the Maryland 400 were encamped with the rest of George Washington's army.

The Army of Ants

The "ants" who "assembled" and "swarmed" in the Mayan myth represent the British army assembling and then attacking the colonialists. This assumes that assembly refers to orderly lines, which real ants aren't known to form. Zipacna's bloody fingernails and hair—which the ants carried in the myth—represent the British army uniforms in red, white, and black.

Zipacna

Scholars indicate that Zipacna carrying the massive pillar—that was too heavy for the 400 Maya men to carry—means Zipacna usurped power. This is consistent with England attempting to usurp power from colonial America. The 400 drunken men in the myth represent the Maryland 400 regiment. The pillar in the myth represents their memorial pillar in Prospect Park, Brooklyn. Allegorically this means four hundred deceased men cannot lift a pillar.

Battle Tactics

In at least one instance, a trick played in the Mayan myth (Zipacna vs the 400 men) represents an historically accurate tactic from this specific moment in the Battle of Brooklyn. For example, the Maryland 400 were tricked into believing they had defeated the British, just as the Maya 400 had been tricked into believing they had defeated Zipacna.

The Old Stone House

The focal point in the Mayan myth was a large house, just as the Maryland 400 and the British were fighting over the "Old Stone House" in Brooklyn. Since the Marylanders were outnumbered five to one—and attacked the British three times—the source of their "drunkenness" was probably fatigue. They "fought like wolves," Lord Cornwallis noted.

THE OLD STONE HOUSE, BROOKLYN, NY, (1800'S).

The Mass Grave

Just as the mythical house fell down upon and killed the Maya 400, so did the Old Stone House symbolically kill the Maryland 400; they died fighting over it. In the myth, the Mayan 400 were buried under the house in a mass grave. Similarly, the Maryland 400 were buried beneath homes in Park Slope, Brooklyn—in a mass grave.

Zipacna's Secret Tunnel

Zipacna's secret escape tunnel in the Mayan myth represents the Park Slope Armory's secret escape tunnel. The armory constructed it long after the Revolutionary War. However, it is located near the Maryland 400's memorial pillar in Prospect Park, just as Zipacna's secret tunnel was located near the Mayan 400's pillar.

The Pillar

Zipacna's victory over the Mayan 400 represents England's victory over the Maryland 400. This story is in the Hall of Records because it changed the course of world history. Accordingly, in the myth, the 400 are honored by being turned into the Pleiades in the afterlife. In Mayan astronomy the Pleiades is the point around which the starry sky rotates. Thus, this represents more allegory. It means the Maryland 400 played a pivotal role. Accordingly, historians (and their pillar) indicate that the Maryland 400 "saved the American army."

A subsequent tale in the Popol Vuh describes how the Hero Twins defeated Zipacna. This represents George Washington's victory over Lord Cornwallis, America's triumph over England, and its subsequent independence.

THE MARYLAND 400'S MEMORIAL PILLAR, BROOKLYN, NY (2015). PHOTO BY AUTHOR.

§

Concealed Wartime Messages

In another story from the Popol Vuh, the lords of Xibalba (England) invited the Hero Twins to visit them. However, fearing a deadly trick, the twins' grandmother hid the message from them by placing it in a gnat. She then hid the gnat in a frog, the frog in a snake, and the snake in a hawk. In the context of the American Revolutionary War this allegorically refers to an America concealing a message from the British. In this instance, fearing a British trick, the American who was supposed to have passed along the message decided against it. This story may or may not have been known to or recorded by historians. Also, since an eagle (represented by a hawk) and a snake indicate an island, this story may have taken place as part of a battle for an island (e.g. the Battle of Brooklyn/Long Island).

Chapter 50

Revolutionary War: Apotheosis

The difference between the Maize God Twins and the Hero Twins is that the latter set could perform miracles. For instance, one brother could kill the other, and the victim could resurrect himself moments later. This represents more allegory. The point is that they cheated death several times and in different ways. Accordingly, George Washington was known as the "bullet-proof president" because several times he had almost been killed (e.g. a bullet tore through his uniform, and a horse had been shot out from under him).

The story of the Hero Twins ends when they fly canoes into the sky and become the sun and the moon. That is allegorical as well. It is an archetype of two rulers, president and vice president. For, just as the moon reflects the light of the sun, so does the vice president reflect presidential policy. Hunapu's deification in the sky represents George Washington's apotheosis (deification). Accordingly, George Washington symbolically resides among the clouds in Brumidi's fresco at the U.S. Capitol. He rests among the clouds, observing the American people below. Thus, the President of the United States replaced Seven Macaw—the King of England—as America's leader "in the sky." Similarly, despite John Adams' ("Xbalanque" in the myth) prediction that history would never honor him, congress recently voted to erect his apotheosis in the National Mall. Hence the apotheoses of George Washington and John Adams are told allegorically as the apotheoses of Hunapu and Xbalanque in Mayan mythology—which is anchored in the Great Ball Court on the map.

THE APOTHEOSIS OF WASHINGTON, BY BRUMIDI.

Having lost the war with America, King George the 3rd's legacy of success was his victory over France in the Seven Years War. "Seven" is the operative word, as Vucub Caquix referred to himself as "Seven Macaw" (Seven Parrot). The name was a source of pride, as macaws are beautiful parrots and Vucub Caquix was worth seven of them. In addition, both King George and the mythical bird had harmful quack doctors and died in obscure poverty.

To summarize the myths, no one can lift a house that sleeps 400 men. Parrots can't hold extensive conversations, and severed heads cannot impregnate anyone. Stories from the Yucatan Hall of Records are symbolic and geographically anchored within the map. The architect somehow knew about George Washington, the Capitol Dome, Brumidi's fresco, King George III etc. a thousand years in advance.

Chapter 51

The American Civil War

Recount of the Revolutionary War

The story of America's founding is told as a creation myth. The colonial period (one of the creation periods) ends with the defeat of the pre-revolutionary founding fathers—the Maize God Twins (as presented in previous chapters). America's transition to independence is told as the rise of the Hero Twins (George Washington and John Adams) and their conflict with the bird-demon Vucub Caquix (King George III) and his son Zipacna (Lord Cornwallis). The creation myth includes Zipacna's victory over the "Four Hundred Drunken Men," (the Maryland 400), and the Hero Twins' ultimate victory over Zipacna and the Lords of Xibalba (England), which included Vucub Caquix and his "sons."

The American Civil War

A subsequent creation myth about "the beginning of civilized life" (Schele and Matthews) tells of two nearby capital cities—"Snake Hill" and "Place of the Cattail Reeds" (or "Tollan"). Snake Hill (or "Coatepec"), located in the South, represents Richmond, VA—capital of the Confederacy, where an actual "Snake Hill" is a prominent landmark near the University of Richmond. The Place of the Cattail Reeds, in the north, represents Washington, D.C.—capital of the Union.

The Confederacy

The actual city of Coatepec is located in Veracruz, Mexico. It is where African slaves were brought during the Transatlantic Slave Trade. It is also where megaliths of African (Olmec) heads were erected over a thousand years prior to the slave trade. An image below shows "El Negro," a muscular, African man, having been subdued. He is laying face-down with his chin on the ground. It is a painful position for his neck, and his eyes are closed, consistent with a man captured for enslavement. As he is tenoned, this means he was inserted into something—presumably a slave ship.

The Olmec statues do not appear in temples or other structures because they are dispossessed—slaves. Thus, the architect used the stone heads to predict African slavery in Coatepec (Veracruz), which he then used metaphorically (in the myth of Huitzilopochtli), to describe Richmond (the Confederacy) as a land of black slaves. As a reminder, the Yucatan Hall of Records cannot be understood apart from Meso-American mythology, which is protective encryption against vandalism and historical distortion. Or, put differently, had the myths been told without encryption, "in plain'English,'" there would be no way to know, currently, if they had been prophetic, or if they had encouraged slavery—meaning, finding the black slave statues somehow triggered the idea of enslaving Africans.

LEFT: OLMEC HEAD, PHOTO BY FURZYK73 | DREAMSTIME. RIGHT: "EL NEGRO" OLMEC HEAD. PHOTO BY JOHN MITCHELL | ALAMY.

RICHMOND DURING THE CIVIL WAR. NOTE: CONFEDERATE CAPITAL BUILDING ON A HILL (CENTER-TOP).

NEWLY-FREED FORMER SLAVES AND DESTROYED BUILDINGS, THE FALL OF RICHMOND, 1865.

The Union

In the myth of Huitzilopochtli, the actual northern "Place of the Cattail Reeds" is unnamed. However, this is consistent with allegory, which minimizes realism. Some experts believe the northern city in the myth was actually Teotihuacan. However, that would place it more *west* of Coatepec, not *north*. The rationale for interpreting the Place of the Cattail Reeds as Washington, D.C. is that the Capitol District was built on and around a former tidal marsh. Cattail reeds presumably grew there, because they still grow in the regional marshes. Moreover, Washington is also near, and north of, Richmond.

The image below shows the U.S. Capitol under construction during the 1860's. The wet section in the lower portion shows the city's canal. It ran east-west to connect the Anacostia River with the Potomac River, both of which contain tidal marshes. In 1791 there were roughly 100 acres of tidal marshlands in and around the main federal buildings (i.e. the White House, etc.) of Washington, D.C. A large part of the marsh formed a semicircle west of, and around, the Capitol Building.

UPPER: THE U.S. CAPITOL BUILDING UNDER CONSTRUCTION.
LOWER: WASHINGTON, D.C. CANAL, 1860'S.

Abraham Lincoln

In the myth, "Huitzilopochtli" was the leader of both the northern and southern capital cities. Evidence that his story is mythological (or allegorical) is that he was born as a full-grown adult, and dressed in combat gear. Huitzilopochtli's military attire refers to Lincoln's role in the American Civil War.

The ball court that Huitzilopochtli placed "at the base" of the Southern capital in the myth refers to Washington, D.C. According to the myth, conflict arose when "the Southerners" (Schele's and Matthews' terminology)—portrayed as an unfaithful woman who becomes pregnant—wanted to secede and create a new nation that involved taking over Mexico City. This refers to the Confederacy's (i.e. the Knights of the Golden Circle) plan to expand beyond the American South into Central America and the Caribbean. All of the indigenous and African people in that region were to be enslaved, with Mexico City being the first major target.

The War

According to Schele and Matthews, the South's behavior "...angered Huitzilopochtli who came down from his mountain armed for war." Then he surrounded the Southerners and defeated them. In one version of the myth, the Southerners kill their mother, who presumably represents the United States. An image shows Huitzilopochtli's (Lincoln's) mythic decapitation of Coyolxaucihuatl—the Southern woman from *Snake* Hill—on the western wall of the Great Ball Court (the Washington, D.C. National Mall replica). Her origin is indicated by her blood in the form of six snakes (see image below). Perhaps this scene is separate from the other decapitation scene at the Great Ball Court because they have different interpretations—political vs geological.

HUITZILOPOCHTLI AFTER DECAPITATING COYOLXAUCIHUATL (AUTHOR'S INTERPRETAION) AT THE GREAT BALL, CHICHEN ITZA. PHOTO BY HJPD.

Allegorically, Huitzilopochtli's (Lincoln's) decapitation of the unfaithful woman at the Great Ball Court (Washington, D.C.), means he defeated the Southern leadership. His subsequent cannibalization of the Southern woman's heart places him at the Platform of Eagles and Jaguars—the White House—where the two animals are shown eating human hearts. Since the Platform of Eagles and Jaguars is the only temple in Chichen Itza's U.S. section where human hearts are eaten, this identifies Huitzilopochtli as the President of the United States.

The reference to "400" Southerners in the myth is based on the archetype of brave soldiers (i.e. the Maryland 400) from the Revolutionary War. Lincoln (Huitzilopochtli) was known as the "God of War and Sacrifice" because he was *sacrificed* (assassinated) at the end of the Civil *War*. The ritual of Meso-American human sacrifices made to Huitzilopochtli represents Civil War casualties. To reiterate, the above myth was only understandable because it was anchored in a specific temple at Chichen Itza—the Great Ball Court. Without the map, the Mayan myths cannot be fully understood.

Modern Washington, D.C. (Tollan)

According to Schele and Matthews, the inhabitants of Tollan (allegorically, Washington, D.C.) were known as "great sages who invented the calendar, divination, astronomy, the arts, writing, medicine, monumental architecture, the institutions of government, agriculture, money, and all things civilized." Some of the parallels with modern Washington, D.C. are apparent—e.g. monuments and statues, government offices, the city's massive solar calendar, the zodiac in the Capitol Building and in other places, the Smithsonian art museum, the Federal Reserve Bank, etc. The myth also describes the people of Cattail Reed city (Washington, D.C.) as rich, pious, jewelry consumers, who maintain the society's standard for "speaking properly." Arguably, most of the above characteristics apply to Washington, D.C.

Constructing the American Government

Schele and Matthews (1998) presented a Mayan "*Hearth of Creation* to center the new order," which consisted of three thrones. Upon them sat a jaguar, a snake, and an alligator/shark who was a representative of Itzamna (the lizard and highest god). The thrones represent the three branches of American government: the Executive Branch (Itzamna), the Judicial Branch (the wise jaguar priest) and the Legislative Branch (snakes symbolize people in Mayan mythology). All are located in or are associated with offices in Federal Triangle. This is based on the Schele and Matthews description of the *hearth of thrones* residing inside the "triangle of stars" below Orion's belt.

The National Mall—geographically the heart of American government—is described allegorically in the Mayan myth as a large, four-sided area, with four corners, and a central "tree" called "Raised-up Sky." According to Schele and Matthews, the tree was "erected" in

February, and represents the "space in which we all live." Since trees are grown rather than erected, and are not grown in any particular month, this tree represents the Washington Monument. The monument is an obelisk that was erected in the center of the National Mall (a four-sided figure) and dedicated on February 21st 1885. The Mayas, however, set the date of erection at February 5th. This is not necessarily an anomaly. It could refer to the date of the monument's completion (i.e. finishing touches) prior to its dedication. Assuming "Raised-up Sky" is analogous to a central tent pole, then the "tree" represents the tallest local structure. Accordingly, by law, no other structure in Washington, D.C. is permitted to exceed the height of the Washington Monument.

THE WASHINGTON MONUMENT.

Chapter 52

The East Coast and the Mississippi Valley

Florida

The massive stone wall running south from the Washington, D.C. replica on the Mayan map represents the Atlantic coast of the United States. It terminates at a replica of Florida. Like the Gulf Coast of the United States, this wall has a right-angle extension that is concaved south. The little extended wall has a groove in it where the Mississippi Valley would begin at the Port of New Orleans. The walls appear to have been designed so that a heavy rain would fill the Gulf of Mexico replica; then the overflow would flow into the groove, thus replicating the Mississippi River. Thus, both the American and European replica sides of the Mayan map are consistent with contemporary maps. In the map image below, note the peninsula's lower and upper elevations.

LEFT: MAYAN MAP OF FLORIDA PENINSULA,
GULF OF MEXICO, AND PANHANDLE.
RIGHT: MODERN MAP OF FLORIDA.

The Americas: The Southwestern U.S. and Old Mexico

Chapter 53

New Mexico

Mayan map structures southwest of Washington D.C. (the Great Ball Court) represent structures in the Southwestern American landscape. Thus, traveling southwest from the Great Ball Court takes us to a replica of Chaco Canyon, New Mexico. It was part of the territory of the Anasazi, the "Ancient Ones." Local Native American myths, here, include a story about "Wing-makers" who visited them from the future, and who set up a type of base camp.

Chaco Canyon is known for several structures, including Fajada ("Banded") Butte, the Great Kiva, pueblos and petroglyphs. Fajada Butte stands 135 meters above the desert floor. When the moon is at its extreme position in its 19 year cycle, a shadow at the butte pierces a spiral petroglyph in half at moonrise. Similarly, when the moon is at its other extreme, its shadow arrives on the left-hand side a larger spiral.

Fajada Butte's replica on the Mayan map is an observatory-temple called "El Caracol" ("the Snail"). It was so-named because of its coiled interior ramp. Both Fajada Butte and El Caracol have coiled ramps. Each also has rings and windows for observation at the top. The New Mexico butte has megaliths that indicate the solar holidays, and the Mayan temple indicates solar holidays as well.

EL CARACOL, CHICHEN ITZA. BY JESUS ELOY RAMOS LARA.

FAJADA BUTTE, CHACO CANYON, NEW MEXICO.

A second distinguishing feature of New Mexico's Chaco Canyon is its Great Kiva. Kivas are circular brick pits that served religious purposes. There is a small replica of a kiva adjacent to El Caracol, just as the Great Kiva is near Fajada Butte. Both the Great Kiva in Chaco Canyon, and its replica on the Mayan map are round with north-south stairs and a keyhole shape. Some guides at Chichen Itza say the tiny kiva was used for viewing eclipses.

GREAT KIVA (KEYHOLE-SHAPED), CHACO CANYON, NEW MEXICO.

MINIATURE KIVA NEXT TO EL CARACOL, CHICHEN ITZA.

EL CARACOL (TOP) AND MINIATURE
KIVA (BOTTOM), CHICHEN ITZA.

Chapter 54

Arizona: The Grand Canyon

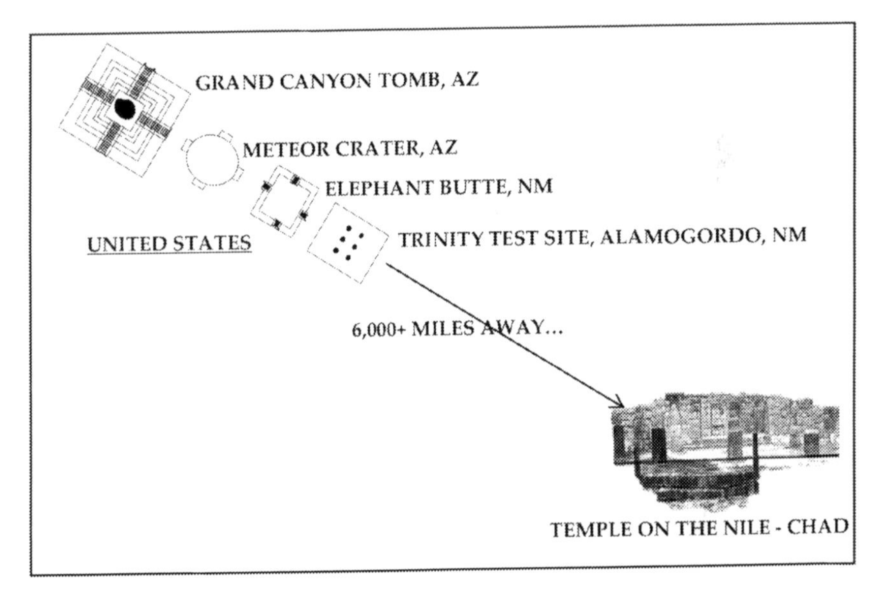

ALIGNMENT: GRAND CANYON, AZ - METEOR CRATER, AZ, ELEPHANT BUTTE, NM – ALAMOGORDO, NM – CHAD, AFRICA.

REPLICA ALIGNMENTS (LEFT TO RIGHT): GRAND CANYON, ELEPHANT BUTTE AND ALAMOGORDO. NOTE: THE METEOR CRATER'S REPLICA IS TOO SHORT TO BE SEEN IN THIS IMAGE.

Traveling west from Chaco Canyon, the next outstanding natural landmarks on both the map and in actuality are the Grand Canyon and the Meteor Crater in Arizona. On the Mayan map the Grand Canyon is called the Tomb of the High Priest (El Osario). It is a hollow pyramid with a shaft that runs 39 feet from the roof to a cave. In other words it represents a mountain with a deep hole in it—a general description of the Grand Canyon.

The cave that connects with the shaft on the Mayan map represents G.E. Kincaid's cave in the Grand Canyon (See Chapter 54 Appendix). Kinkaid said he found it overlooking the Colorado River. The April 5th 1909 Arizona Gazette described the cave as containing mummies covered in clay and bark, tablets engraved with hieroglyphics, copper urns, broken swords, cooking vessels, an image of a prehistoric animal, many rooms, a room that smelled "snaky," and a "shrine." The shrine had yellow stones and a statue of a man sitting cross-legged and holding lotus flowers.

Archaeologist Edward H. Thompson reported jade beads, skeletons, and other artifacts inside the Chichen Itza cave (the replica of Kinkaid's cave in Arizona). Presumably no priest was found there because the tomb simply replicates Kincaid's Tomb in the Grand Canyon.

The upper façade of the Tomb of the High Priest once contained stone carvings of roughly half a dozen ceremonial dancers. Each wore a headdress of feathers, a bird-face mask, and many decorative feathers. Each carried a stick/scepter in one hand and a drum in the other. According to archaeologist Peter Schmidt, one dancer was "captured at the moment of executing a whirling dance." The evidence of the twirl is that one of the statue's arms is flush with the wall, while the other extends out from it. Cloud volutes place them in the sky.

The dance, which is still performed that way, represents the Grand Canyon's Hopi Indians' "Eagle Dance." Also known as Thunder birds, ancient Native Americans revered eagles, and believed they carried messages between heaven and earth. Thus, the dancers' outstretched arms represent soaring messengers. Presumably, then, Kinkaid found the tomb of an important Hopi shaman in the Grand Canyon, as the Mayan map of the Grand Canyon indicates.

There has been much debate over the authenticity of Kinkaid's claim—the Smithsonian has no records of him or "Professor Jordan." However, it seems unlikely that mere chance placed a religious leader in a canyon cave on the map and in reality, and in the same geographic locations. Thus, the Mayan map appears to corroborate Kinkaid's story. Perhaps more importantly, the map includes landmarks (e.g. a mammoth grave, the meteor crater, Elephant Butte, etc.) that indicate the location of Kinkaid's discovery. The key to pinpointing the cave's location may be the shadow of the adjacent Mastodon Pillar (discussed in a later chapter) on a certain day of the year, at a certain time of day. Finally, the "Owl Man," farther below, is a well-known, demon-like icon among the Apache and other tribes in the Grand Canyon region.

TOMB OF THE HIGH PRIEST, CHICHEN ITZA.
PHOTO BY AUTHOR.

DANCER AT THE TOMB OF THE HIGH PRIEST.
PHOTO BY JOEL SKIDMORE. USED WITH PERMISSION.

THE GRAND CANYON, BY THE NATIONAL PARK SERVICE.

SHAFT OF THE TOMB OF THE HIGH PRIEST. USED WITH PERMISSION. CARNEGIE INSTITUTION.

OWL MAN, TOMB OF THE HIGH PRIEST. USED WITH PERMISSION. CARNEGIE INSTITUTION.

Chapter 54 Appendix
(The Kinkaid Article)

EXPLORATIONS IN GRAND CANYON:

THE ARIZONA GAZETTE of April 5th, 1909

**Mysteries of Immense High
Cavern Being Brought
to Light**

JORDAN IS ENTHUSED

**Remarkable Finds Indicate
Ancient People Migrated
From Orient**

The latest news of the progress of the explorations of what is now regarded by scientists as not only the oldest archeological discovery in the United States, but one of the most valuable in the world, which was mentioned some time ago in the Gazette, was brought to the city yesterday by G.E. Kinkaid, the explorer who found the great underground citadel of the Grand Canyon during a trip from Green River, Wyoming, down the Colorado, in a wooden boat, to Yuma, several months ago.

According to the story related to the Gazette by Mr. Kinkaid, the archeologists of the Smithsonian Institute [1], which is financing the expeditions, have made discoveries which almost conclusively prove that the race which inhabited this mysterious cavern, hewn in solid rock by human hands, was of oriental origin, possibly from Egypt, tracing back to Ramses. If their theories are borne out by the translation of the tablets engraved with hieroglyphics, the mystery of the prehistoric peoples of North America, their ancient arts, who they were and whence they came, will be solved. Egypt and the Nile, and Arizona and the Colorado will be linked by a historical chain running back to ages which staggers the wildest fancy of the fictionist.

A Thorough Examination

Under the direction of Prof. S. A. Jordan, the Smithsonian Institute is now prosecuting the most thorough explorations, which will be continued until the last link in the chain is forged. Nearly a mile underground, about 1480 feet below the surface, the long main passage has been delved into, to find another mammoth chamber from which radiates scores of passageways, like the spokes of a wheel.

Several hundred rooms have been discovered, reached by passageways running from the main passage, one of them having been explored for 854 feet and another 634 feet. The recent finds include articles which have never been known as native to this country, and doubtless they had their origin in the orient. War weapons, copper instruments, sharp-edged and hard as steel, indicate the high state of civilization reached by these strange people. So interested have the scientists become that preparations are being made to equip the camp for extensive studies, and the force will be increased to thirty or forty persons.

Mr. Kinkaid's Report

Mr. Kinkaid was the first white child born in Idaho and has been an explorer and hunter all his life, thirty years having been in the service of the Smithsonian Institute. Even briefly recounted, his history sounds fabulous, almost grotesque.

"First, I would impress that the cavern is nearly inaccessible. The entrance is 1,486 feet down the sheer canyon wall. It is located on government land and no visitor will be allowed there under penalty of trespass. The scientists wish to work unmolested, without fear of archeological discoveries being disturbed by curio or relic hunters.

A trip there would be fruitless, and the visitor would be sent on his way. The story of how I found the cavern has been related, but in a paragraph: I was journeying down the Colorado river in a boat, alone, looking for mineral. Some forty-two miles up the river from the El Tovar Crystal canyon, I saw on the east wall, stains in the sedimentary formation about 2,000 feet above the river bed. There was no trail to this point, but I finally reached it with great difficulty.

Above a shelf which hid it from view from the river, was the mouth of the cave. There are steps leading from this entrance some thirty yards to what was, at the time the cavern was inhabited, the level of the river. When I saw the chisel marks on the wall inside the entrance, I became interested, securing my gun and went in. During that trip I went back several hundred feet along the main passage till I came to the crypt in which I discovered the mummies. One of these I stood up and photographed by flashlight. I gathered a number of relics, which I carried down the Colorado to Yuma, from whence I shipped them to Washington with details of the discovery. Following this, the explorations were undertaken.

The Passages

"The main passageway is about 12 feet wide, narrowing to nine feet toward the farther end. About 57 feet from the entrance, the first side-passages branch off to the right and left, along which, on both sides, are a number of rooms about the size of ordinary living rooms of today, though some are 30 by 40 feet square. These are entered by oval-shaped doors and are ventilated by round air spaces through the walls into the passages. The walls are about three feet six inches in thickness.

The passages are chiseled or hewn as straight as could be laid out by an engineer. The ceilings of many of the rooms converge to a center. The side-passages near the entrance run at a sharp angle from the main hall, but toward the rear they gradually reach a right angle in direction.

The Shrine

"Over a hundred feet from the entrance is the cross-hall, several hundred feet long, in which are found the idol, or image, of the people's god, sitting cross-legged, with a lotus flower or lily in each hand. The cast of the face is oriental, and the carving this cavern. The idol almost resembles Buddha, though the scientists are not certain as to what religious worship it represents. Taking into consideration everything found thus far, it is possible that this worship most resembles the ancient people of Tibet.

Surrounding this idol are smaller images, some very beautiful in form; others crooked-necked and distorted shapes, symbolical, probably, of good and evil. There are two large cactus with protruding arms, one on each side of the dais on which the god squats. All this is carved out of hard rock resembling marble. In the opposite corner of this cross-hall were found tools of all descriptions, made of copper. These people undoubtedly knew the lost art of hardening this metal, which has been sought by chemicals for centuries without result. On a bench running around the workroom was some charcoal and other material probably used in the process. There is also slag and stuff similar to matte, showing that these ancients smelted ores, but so far no trace of where or how this was done has been discovered, nor the origin of the ore.

"Among the other finds are vases or urns and cups of copper and gold, made very artistic in design. The pottery work includes enameled ware and glazed vessels. Another passageway leads to granaries such as are found in the oriental temples. They contain seeds of various kinds. One very large storehouse has not yet been entered, as it is twelve feet high and can be reached only from above. Two copper hooks extend on the edge, which indicates that some sort of ladder was attached. These granaries are rounded, as the materials of which they are constructed, I think, is a very hard cement. A gray metal is also found in this cavern, which puzzles the scientists, for its identity has not been established. It resembles platinum. Strewn promiscuously over the floor everywhere are what people call "cats eyes', a yellow stone of no great value. Each one is engraved with the head of the Malay type.

The Hieroglyphics

"On all the urns, or walls over doorways, and tablets of stone which were found by the image are the mysterious hieroglyphics, the key to which the Smithsonian Institute hopes yet to discover. The engraving on the tables probably has something to do with the religion of the people. Similar hieroglyphics have been found in southern Arizona. Among the pictorial writings, only two animals are found. One is of prehistoric type.

The Crypt

"The tomb or crypt in which the mummies were found is one of the largest of the chambers, the walls slanting back at an angle of about 35 degrees. On these are tiers of mummies, each one occupying a separate hewn shelf. At the head of each is a small bench, on which is found copper cups and pieces of broken swords. Some of the mummies are covered with clay, and all are wrapped in a bark fabric.

The urns or cups on the lower tiers are crude, while as the higher shelves are reached, the urns are finer in design, showing a later stage of civilization. It is worthy of note that all the mummies examined so far have proved to be male, no children or females being buried here. This leads to the belief that this exterior section was the warriors' barracks.

"Among the discoveries no bones of animals have been found, no skins, no clothing, no bedding. Many of the rooms are bare but for water vessels. One room, about 40 by 700 feet, was probably the main dining hall, for cooking utensils are found here. What these people lived on is a problem, though it is presumed that they came south in the winter and farmed in the valleys, going back north in the summer.

Upwards of 50,000 people could have lived in the caverns comfortably. One theory is that the present Indian tribes found in Arizona are descendants of the serfs or slaves of the people which inhabited the cave. Undoubtedly a good many thousands of years before the Christian era, a people lived here which reached a high stage of civilization. The chronology of human history is full of gaps. Professor Jordan is much enthused over the discoveries and believes that the find will prove of incalculable value in archeological work.

"One thing I have not spoken of, may be of interest. There is one chamber of the passageway to which is not ventilated, and when we approached it a deadly, snaky smell struck us. Our light would not penetrate the gloom, and until stronger ones are available we will not know what the chamber contains. Some say snakes, but other boo-hoo this idea and think it may contain a deadly gas or chemicals used by the ancients. No sounds are heard, but it smells snaky just the same. The whole underground installation gives one of shaky nerves the creeps. The gloom is like a weight on one's shoulders, and our flashlights and candles only make the darkness blacker. Imagination can revel in conjectures and ungodly daydreams back through the ages that have elapsed till the mind reels dizzily in space."

An Indian Legend

In connection with this story, it is notable that among the Hopi Indians the tradition is told that their ancestors once lived in an underworld in the Grand Canyon till dissension arose between the good and the bad, the people of one heart and the people of two hearts. Machetto, who was their chief, counseled them to leave the underworld, but there was no way out. The chief then caused a tree to grow up and pierce the roof of the underworld, and then the people of one heart climbed out. They tarried by Paisisvai (Red River), which is the Colorado, and grew grain and corn. They sent out a message to the Temple of the Sun, asking the blessing of peace, good will and rain for people of one heart. That messenger never returned, but today at the Hopi villages at sundown can be seen the old men of the tribe out on the housetops gazing toward the sun, looking for the messenger. When he returns, their lands and ancient dwelling place will be restored to them. That is the tradition.

Among the engravings of animals in the cave is seen the image of a heart over the spot where it is located. The legend was learned by W.E. Rollins, the artist, during a year spent with the Hopi Indians.

There are two theories of the origin of the Egyptians. One is that they came from Asia; another that the racial cradle was in the upper Nile region. Heeren, an Egyptologist, believed in the Indian origin of the Egyptians. The discoveries in the Grand Canyon may throw further light on human evolution and prehistoric ages.

Chapter 55

Arizona: The Meteor Crater and Elephant Butte

The Meteor Crater

The Meteor Crater in Winslow Arizona is the largest intact crater in the world. It is one mile in diameter, 2.4 miles in circumference, and 550 ft. deep. This enormous circle is located southeast of the nearby Grand Canyon, and is replicated on the Mayan map by a stone circle.

Located roughly 30 meters southeast of the Grand Canyon replica (Tomb of the High Priest), the meteor crater replica is roughly 15 ft. in diameter and about 1 ft. high. It is normally filled with dirt and grass, but it may have once been empty, like a meteor crater.

The actual Arizona Meteor Crater was divided into four land claim quadrants, referred to as "Venus," "Mars," "Jupiter" and "Saturn." These match the four markers at Chichen Itza's Meteor Crater replica, which are aligned with northwest, northeast, etc. Thus, the architect may have included some of the crater's history. Since this structure is the only one of its kind at Chichen Itza, it is not likely that it was placed in the same juxtaposition as is the Meteor Crater with the Grand Canyon by chance.

Elephant Butte

The next structure in this alignment is the Platform of Venus-South. Its identical shape with the platforms discussed earlier mean it represents an island—specifically Elephant Island. It is an outstanding landmark because islands are rare in the American desert. One of the best specimens of a stego-mastodon was found on the island.

Archaeologists indicate an alignment between the Tomb of the High Priest (Grand Canyon replica), the unnamed circle (the meteor crater replica), the Platform of Venus (Elephant Butte), a "Funerary Structure" (Alamogordo, NM) and the Temple of Xtoloc near the southern Cenote of Xtoloc (Lake Chad). This alignment is consistent with a quote from the Kinkaid article.

"Egypt and the Nile, and Arizona and the Colorado will be linked by a historical chain running back to ages which staggers the wildest fancy of the fictionist." (G.E. Kinkaid).

The alignment ends at the replica of the center of Africa's Sahelian Belt. It is an area historically known for devastating droughts that may last for centuries.

METEOR CRATER, ARIZONA.

METEOR CRATER REPLICA, CHICHEN ITZA.

PLATFORM OF VENUS-SOUTH (ELEPHANT ISLAND REPLICA).

Chapter 56

Alamogordo and Pakal

The next structure in the alignment is the "Funerary Structure." It represents the observation bunker in which Robert Oppenheimer, other scientists, and military officers observed the world's first nuclear explosion at Trinity Test Site in Alamogordo, NM. The structure's name is appropriate, as Trinity is located in a desert called "A Single Day's Journey of Death." On the Mayan map this structure is located along the alignment where Alamogordo would be found.

LEFT: FUNERARY STRUCTURE AT CHICHEN ITZA.
RIGHT: TRINITY TEST BASE CAMP, ALAMOGORDO, NM.

LEFT: FIREBALL AT THE TRINITY NUCLEAR TEST SITE. RIGHT: ROBERT OPPENHEIMER AND BRIGADIER GENERAL LESLIE GROVES AT THE REMAINS OF TRINITY TEST SITE (1945).

PILLAR OF MASTODONS.

A few meters south of the Tomb of the High Priest stands a pillar of three stacked heads. With their large ears and elephant-like noses, they seem to represent mastodons, mammoths, or gomtheres. Maude Blackwell, Alice Le Plongeon's friend who inherited her photos, referred to them as mastodons. Thus, the pillar may represent a graveyard of early mastodons.

Pakal (Pacal)

Somewhere between New Mexico and Arizona the map includes a stele which displays a date in AD 870. The stele shows a hornbill duck and mentions "Pakal." As explained previously, that name appears to mean "space traveler," "captain," "flying traveler," etc. Since there are no hornbill ducks in the Southwest United States, and presumably never were, then the duck's presence on the stele may symbolize travel to the Arizona region from elsewhere. Perhaps the Pakal on the stele refers to the Wingmakers (time and space travelers) of local, Native American mythology. The date (AD 870) may indicate when the travelers arrived in the American desert. The tablet is on display in the Gran Museo del Mundo Maya in Merida.

Chapter 57

The Anasazi

The ancient Anasazi of the Southwestern United States often lived in cliff dwellings. The most popular among these is at Mesa Verde, Colorado. However, there are many of such structures in the area. Ancient cliff dwelling areas commonly include petroglyphs, some of which are images of human hands.

The architect presented the cliff dwellings near Sedona, AZ as a structure called Akab Dzib, which means "House of Mysterious Writing." Chichen Itza's nearby House of Colorado describes Akab Dzib as "long and flat with an excessive number of chambers." Akab Dzib was constructed at the edge of a steep, dry, cenote to replicate a desert cliff dwelling. The enigmatic red handprints inside the house and the name "mysterious writing," refer to the red earth petroglyphs of the Anasazi. The structure's shape—long with a low ceiling—is consistent with the general appearance of cliff dwellings.

The "House of Colorado" (discussed in the next section) mentions Akab Dzib because the cliff dwellings are located in the Colorado Plateau (Arizona, Colorado, etc.). Local legends say fairy-like creatures lived in Akab Dzib long ago, and that the glyphs glow at midnight. Perhaps the illuminated glyphs refer to the glow-in-the dark, cereus flower in the Arizona desert.

LEFT: AKAB DZIB (SOUTHWEST CORNER LOOKING NORTHEAST). PHOTO BY AUTHOR. RIGHT: NIGHT-BLOOMING CERUS FLOWER BY ROBERT THORNTON (1799).

Chapter 58

Mestizos & Corn

The next structures southwest of the Grand Canyon replica on the Mayan map are the "House of the Corn Grinders" and the "House of the Mestizos." Native Americans and Mestizos have been grinding corn to make flour for tacos, tortillas, etc. for centuries. Mestizos are an ethnic blend of Native Americans and Spaniards. They are known as Mexicans in what is now the Southwest United States—where the "House of the Mestizos" is located on the Mayan map (called "Mexican Homelands").

It is unlikely that, by mere chance, the architect placed Mexicans in the Southwest United States (Old Mexico) and just happened to say they eat ground corn products. Rather, he or she knew where the old U.S.-Mexican (Mestizo) border was, even though neither the U.S. nor Mestizos existed when the Chichen Itza map was built. He knew Mestizos would exist in the future, where they would live, and what they would eat.

BOTTOM: HOUSE OF THE CORN-GRINDERS AND HOUSE OF THE MESTIZOS. TOP: TOMB OF THE HIGH PRIEST.

Chapter 59

The Colorado Plateau

A short distance southwest of the House of the Mestizos (i.e. Mexican Homelands) and the House of the Corn Grinders (i.e. people who eat ground corn foods), we find the "House of Colorado" (Home of Red Rocks). It refers to the Colorado Plateau, which is comprised of parts of Colorado, New Mexico, Utah and Arizona—and home of red rocks. The four states meet at the "Four Corners" landmark.

Consistent with the Colorado Plateau's high elevation (e.g. Denver, Colorado is a mile high), the House of Colorado stands on a plateau-like foundation, which represents a society on a plateau. The ball court on the east side of the House of Colorado represents the Rio Grande Rift—a north-south fault line on the east side of the Colorado Plateau.

HOUSE OF COLORADO
(HOME OF RED ROCKS), CHICHEN ITZA.

REPLICA OF RIO GRANDE RIFT AT "HOUSE OF COLORADO."

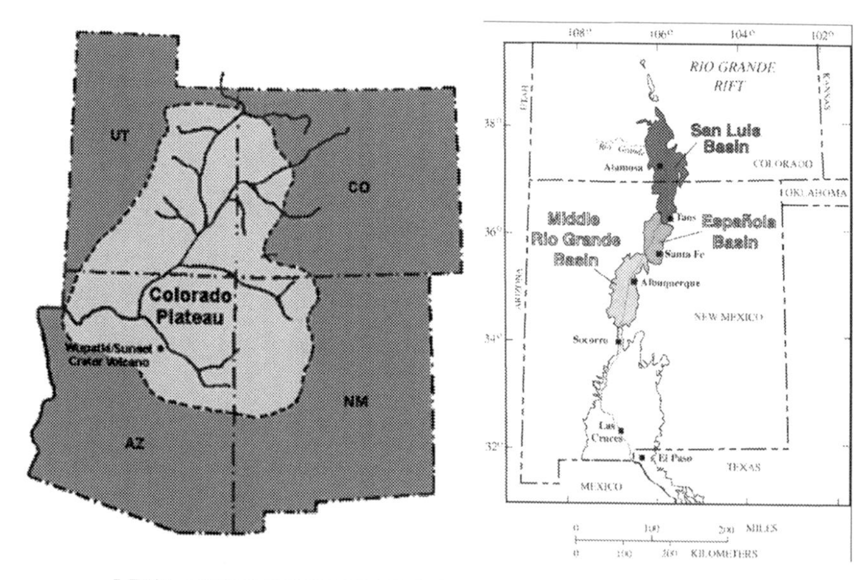

LEFT: MAP OF COLORADO PLATEAU. RIGHT: MAP OF RIO GRANDE RIFT IN COLORADO AND NEW MEXICO.

The House of Colorado was not randomly placed on a man-made plateau or located where we find the Colorado Plateau on contemporary maps. Consider the big picture: Mexican and Native American peoples who eat ground-corn foods (e.g. tortillas), the Grand Canyon (Arizona), the Meteor Crater (Arizona), Fajada Butte and the kiva (Chaco Canyon, New Mexico), the Colorado Plateau (Colorado, Arizona, etc.).

The Rio Grande Rift

The ball court adjacent to the east side of the House of Colorado represents the Rio Grande Rift. The decapitation scene at the ball court shows two men falling from their feet simultaneously, as though they were suddenly caught by surprise. The image is different from the one at the Great Ball Court, which is clearly an execution scene. If the map theory is accurate, then a previous or future destructive earthquake on the Rio Grande Rift was/will come with little warning.

RIO GRANDE GORGE IN RIO GRANDE RIFT.

California

Immediately west of the House of Colorado is the "House of the Deer." It is at the western border of Chichen Itza. Some sources indicate that deer symbolized "west" to the Mayas. Thus, this structure indicates the western part of the map, presumably the California region.

Chapter 60

The Spanish Mission Cities

As we continue southwest on the Chichen Itza map, we arrive at "La Iglesia" and "the Nunnery." They represent the Spanish Mission cities—Las Cruces ("The Crosses"), San Diego ("Saint James"), Los Angeles ("The Angels"), etc.—of today's Southwestern United States and Northern Mexico.

The Mayan map predicted the recent (drug-related) beheadings in Northern Mexico long ago (See image below). Since the researchers who supplied the photos of human heads did not mention their bodies, apparently the heads were constructed and displayed alone. Moreover, it is unusual for the eyes to be closed on Mayan statues, as we see among some of the heads here. Thus, the image shows the aftermath of a mass-execution.

DECAPITATIONS IN THE SPANISH MISSION CITIES.

THE NUNNERY AND THE CHURCH, 1900, BY LE PLONGEON. CORNELL UNIVERSITY LIBRARY.

THE NUNNERY AND THE CHURCH (“LA IGLESIA”) - SPANISH MISSION CITY REPLICAS. PHOTO BY AUTHOR.

The Americas: Central and South America

Chapter 61

Central and South America: Old Chichen

The main discussion of the map ends here, because the Central and South America portion of the map ("Old Chichen") has been closed since I began developing the map theory in 2012. Therefore, I have not yet observed it. However, below I offer a few observations.

The "House of the Paneled Walls" may represent either Mayas or places of Mayan influence. The rationale is that several of the Mayan cities (e.g. Chichen Itza and Palenque) feature paneled walls. The "House of the Monkeys" may represent the place where—as we travel south—we begin to see monkeys. This would begin in Mexico and continue into Central America and beyond. The "House of the Owl" may represent the "Owl Man" image on the Nazca Plains in Peru.

The "House of the Phalli" may represent the South American tribesmen who wore gourds as penis sheaths. There is a Stonehenge-type structure in Old Chichen, which may represent the "Calçoene Megalithic Observatory" ("Amazon Stonehenge"). It consists of 127 granite blocks (some up to 4 meters tall) in a 30 meter diameter circle that is believed to have been an astronomical and burial site.

Finally, the "Platform of the Great Turtle" may represent Easter Island, which resembles a large turtle. The stone heads on the island only appear where a turtle's shell would be. Thus, they do not appear at the head or the tail.

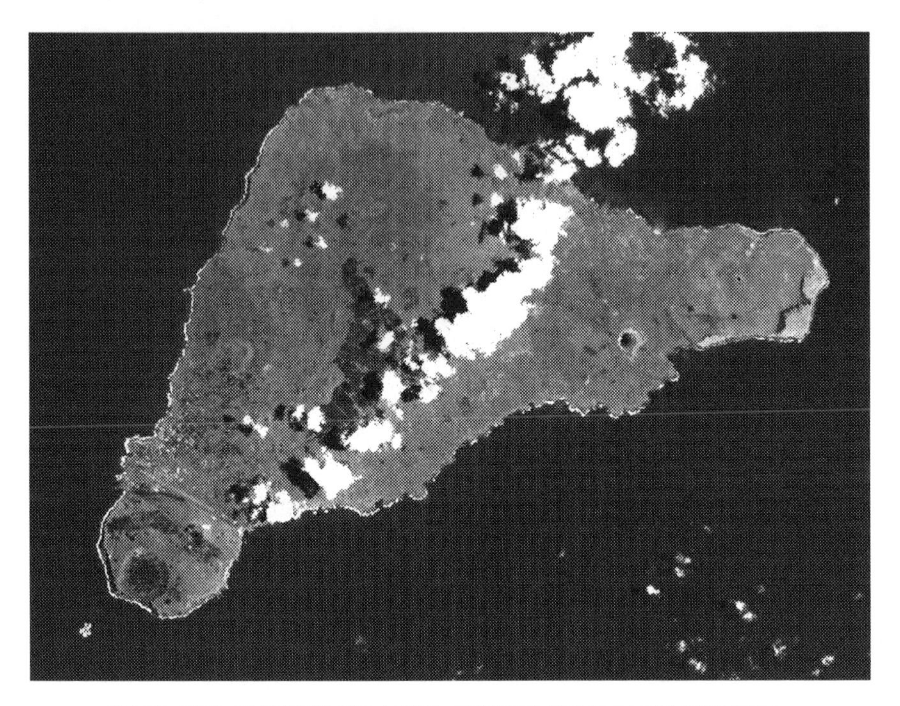

RAPA NUI (EASTER ISLAND). SATELLITE IMAGE BY NASA.

Chapter 62

Summary

This discussion theorized that the legendary Yucatan Hall of Records is mainly comprised of rituals and myths anchored in temples that represent planets on a universal map. Chichen Itza is a map of the Western Hemisphere. Thus, the myths and rituals connected with Chichen Itza help tell the story of the Western Hemisphere. Similarly, any myths or rituals at Uxmal help tell the story of the Eastern Hemisphere. If the map theory is correct, then the Hall of Records tells the stories of roughly 4,000 planets with intelligent life, based on the estimated number of ancient Maya sites that professors Witschey and Brown identified. The map theory hypothesizes that a time-traveler "architect" designed the Hall of Records.

The architect's identify remains unknown. However, he or she may have been an American military member of a secret society. Consider the following:

- America's two major encounters with UFO's (Area 51 and Rendlesham Forest) happened at American air bases, roughly on the 4th of July and Christmas (Two important holidays for Americans).
- The U.S. Air Military does not generally seem to aid UFO investigations.
- The architect offered many details about Washington, D.C. and American history.
- The pivotal Mayan myths (i.e. the "Maize God Twins" and the "Hero Twins") match the narrative of early America.
- The Maya's foundational myth about the Maize Gods Twins matches the role of maize in the founding of the new masonic "lodge" that is America.

- It may not be coincidental that Freemasons revere the "Great Architect of the Universe," that the Mayan "architect" created a replica of the universe, and that Augustus Le Plongeon—a high-ranking Freemason—went to the Yucatan to find the origins of Freemasonry.
- There are thousand year old images of Secret Service soldiers—with modern military equipment—on the roof of the White House replica.
- The architect identified a U.S. Airbase in Iceland, and placed over fifteen dozen replicas of nuclear weapons and symbols of military pilots there.
- There are images of white men at Chichen Itza, even though it was built 500 years before any known white men (the Spanish Conquistadors) arrived there.
- The Mayan map included Normandy Beach and the story of the German invasion of France in World War II.
- The architect replicated the Vietnam War Memorial in the Washington, D.C. replica.
- When deciphered (re: the three-part puzzle), the sarcophagus of Lord Pakal presents him as an American astronaut.
- It is probably not by coincidence that American Air Force identification is worn on an embroidered patch called a "shield," and that "Pakal" means "shield" in the ancient Maya language.
- The Mayas' highest level of heaven ("Omeyocan") sounds suspiciously like "American" when pronounced with a foreign (i.e. ancient Maya) accent or dialect.
- The story of Huitzilopochtli and his fight with the "Southerners" is similar to the story of the American Civil War.

However implausible time travel is, Occam's razor is satisfied, and fewer questions remain, if the architect was an American military time traveler, perhaps with ties to a secret society (e.g. "the Illuminati"—the "Enlightened Ones"). If the map theory is valid, then the Yucatan Hall of Records may offer important information about extraterrestrial life.

Index

B

C

D

E

F

G

H

I

J

K

L

M

N

O

P

Q

R

S

X

Y

Z

Printed in the United States
By Bookmasters